Introduction to Maintenance Plant Breeding

The Authors

Dr Phundan Singh graduated in Agriculture from Agra University, Agra and obtained his Master and Doctorate degrees from Kanpur University, Kanpur in Genetics and Plant Breeding. Dr P. Singh has total experience of more than 40 years in the field of Plant Breeding and Genetics. He has served at the Central Institute for Cotton Research [CICR], Nagpur for about 32 years in various capacities such as Scientist, Senior Scientist, Principal Scientist, Head of Division and Director. He has 300 publications to his credit. He has authored 60 books, 18 technical bulletins and contributed 14 chapters in various books. He is an Expert member of Selection Committees and M. Sc. and Ph. D. Examiner in different Agricultural Universities. He received lifetime achievements award from CRDA in 2018. He has visited several countries such as Russia, Belarus, Ukraine, Crimea, Uzbekistan, Canada and USA.

Dr Mridula Billore obtained M. Sc. and Ph.D. degrees from JNKVV, Jabalpur in Genetics and Plant Breeding. Dr Billore has more than 30 year experience of teaching, research and extension and served at JNKVV, Jabalpur/RVSKVV, Gwalior as Assistant Professor, Associate Professor and Dean College of Agriculture. Presently, she is serving as the Dean Faculty at the RVSKVV, Gwalior. She has published more than 70 research papers and 1 book. She received best voluntary pulse research centre award of ICAR in 2010.

Dr D. K. Mishra graduated and post-graduated from Banaras Hindu University, Varanasi and obtained his Ph.D. degree in Genetics and Plant Breeding from JNKVV, Jabalpur. Dr Mishra has more than 40 year experience of teaching, research and extension. He served at JNKVV, Jabalpur in various capacities such as Assistant Professor, Associate Professor, Professor and Head, Dean and Director Farms. He has published more than 100 research papers and 50 extension bulletins, 5 books and 10 manuals.

Introduction to Maintenance Plant Breeding

Dr Phundan Singh

Dr [Mrs.] Mridula Billore

Dr D. K. Mishra

2019

Daya Publishing House®

A Division of

Astral International Pvt. Ltd.

New Delhi – 110 002

ISBN: 9789389569209 (HB)

Published by : **Daya Publishing House®**
A Division of
Astral International Pvt. Ltd.
– ISO 9001:2015 Certified Company –
4736/23, Ansari Road, Darya Ganj
New Delhi-110 002
Ph. 011-43549197, 23278134
E-mail: info@astralint.com
Website: www.astralint.com

Digitally Printed at : **Replika Press Pvt. Ltd.**

Dedicated to following Family members of Senior Author

Mrs. Jaswanti Singh [Wife]
Ms. Pavitra Choudhury [Daughter]
Dr. Saksham Singh [Daughter in-law]
Ms. Shalini Singh [Daughter in-law]

Preface

Varietal maintenance also known as maintenance breeding is an important area of plant breeding. This aspect is offered at postgraduate level in all Indian Stare and Central Agricultural Universities. The information on this important area of plant breeding is not available from any single source. Thus students have to go through huge literature to meet their course requirements. The prime objective of writing this book is to fill this gap and fulfill long felt need of Indian students of plant breeding and genetics. Main features of this book are as follows:

1. This volume contains 18 Chapters related to maintenance of varietal purity. It deals with basics of maintenance breeding, causes of varietal deterioration, categories of improved seed, principles of quality seed production; and production of nucleus seed of cereals, millets, oilseeds, pulses, commercial crops, fibre crops, A, B and R lines and Intellectual Property Rights.
2. Glossary of technical terms has been provided at the end for ready reference of readers.
3. Three appendices, *viz.* List of Agricultural Universities, List of ICAR Research Institutes and List of International Agricultural Research Centers have been provided for benefit of students.
4. Key References have been provided at the end to gather further details about the subject matter.

The cooperation extended by my wife Mrs. Jaswanti Singh during preparation of this book is praiseworthy. Hope, this volume would be useful to the students, researchers, and teachers engaged in the field of Plant Breeding, Genetics and Seed Technology. Constructive suggestions of students and teachers are invited for further improvement of this book.

Phundan Singh

Senior Author

Contents

Post Graduate Syllabus

[As per 5th Deans' Committee Report of ICAR]

PAPER GP 515: MAINTENANCE BREEDING AND CONCEPTS OF VARIETY RELEASE AND SEED PRODUCTION

☆ **UNIT I:** Variety development and maintenance; Definition- variety, cultivar, extant variety, essentially derived variety, independently derived variety, reference variety, farmers' variety, hybrid, and population; Variety testing, release and notification systems in India and abroad.

☆ **UNIT II:** DUS testing- DUS Descriptors for major crops; Genetic purity concept and maintenance breeding.

☆ **UNIT III:** Factors responsible for genetic deterioration of varieties - safeguards during seed production; Maintenance of varieties in self and cross-pollination crops- isolation distance; Principles of seed production; Methods of nucleus and breeder seed production.

☆ **UNIT IV:** Generation system of seed multiplication -nucleus, breeders, foundation, certified, - Quality seed production technology of self and cross-pollinated crop varieties *viz.* Cereals and Millet (wheat, barley, paddy, pearl millet, sorghum, maize, *etc.*); Pulses (green gram, black gram, pigeon pea, chickpea, field pea, lentil); Oilseeds (groundnut, soybean, sesame, castor, sunflower, safflower, linseed, rapeseed and mustard); Seed certification procedures; Seed laws and plant variety protection regulations in India and international systems.

Part I

Introductory Chapters

Chapter 1

Variety Release Procedure

Introduction

A variety refers to a genotype or group of genotypes which have been released for commercial cultivation either by State Variety Release Committee (SVRC) or Central Variety Release Committee (CVRC) and notified by the government of India. A variety is also termed as cultivar (cultivated Variety). A variety has three important characteristics. *viz.* (1) distinctiveness (2) uniformity, and (3) stability. The distinctness means that the variety should be distinguishable by one or more morphological, physiological and biochemical features from all pre-existing cultivars. Uniformity means that the variety should be pure and look uniform, stability means that the variety should give stable performance in different generations. In other words, the variety should remain unchanged in its distinctness and uniformity for reasonable period of time when reproduced or reconstituted (hybrid).

Norms for Release of New Variety

There are certain norms for release of new variety that have to be strictly followed. Besides distinctness, uniformity and stability, the new variety should be superior to the previously released varieties of a crop in one or more of the following characteristics:

1. Yield potential of grain, oil, fibre, fodder, vegetable or other economic product.
2. Resistance to biotic (diseases, insects and parasitic weeds) and abiotic (drought, salinity, frost, cold, heat, metal toxicity, *etc.*) stresses.
3. Quality of oil, fibre, protein, fodder, vegetable, grain, *etc.*
4. Maturity duration (earliness).
5. Adaptability: Suitability for general cultivation over a wide range of environmental conditions.
6. Suitability for machine harvesting, *etc.*

Types of Cultivars

In plant breeding various types of cultivars are developed and released for commercial cultivation. The development of new variety depends mainly on two factors, *viz.* (1) mode of reproduction of the crop species, and (2) method of breeding used for development of a variety. Some varieties are developed by selection without hybridization such as pure line variety, mass selected variety and clonal variety. Development of some varieties involves hybridization and selection. This includes varieties that are developed by pedigree, bulk and backcross methods including multiline cultivars. After release, all these varieties remain under cultivation for several years. The seed of some varieties has to be produced every year afresh such as hybrid varieties and triploids in seed propagated species (*i.e.* watermelon). Some varieties are developed by special breeding techniques such as mutant and transgenic varieties. Different types of varieties which are developed in plant breeding include pure line variety, mass selected variety, clonal variety, synthetic variety, composite variety, multi-line variety, mutant variety, transgenic variety, hybrid variety. These are explained as follows:

Pure Line Variety

It is developed in self-pollinated crops by pure line selection. The progeny of a single homozygous plant obtained by selfing is called pure line. The variety is released from a single best progeny. The base population, which is used for development of pure line variety is always a heterogeneous population. A pure line variety is a homozygous and homogeneous population.

Mass Selected Variety

It is developed in both self and cross pollinated species. However, mass selection is more common in cross pollinated crops than in self-pollinated ones. In self-pollinated crops, a mass selected variety consists of several pure lines. In other words, it is a mixture of several pure lines. Thus it is homozygous but heterogeneous population. In cross pollinated species, a mass selected variety is a mixture of several homo and heterozygotes. Thus it is a heterozygous and heterogeneous population. A mass selected variety has broad genetic base and wide adaptation.

Clonal Variety

It is developed in asexually propagated crops such as potato, sugarcane, sweet potato, *etc.* The progeny of a single best clone is released as a cultivar. Such varieties are heterozygous but homogeneous, because all the plants of a clone have exactly the same genotype. Clonal cultivars have wide adaptation than pure line cultivars. A heterogeneous base population is used for development of clonal variety.

Synthetic Variety

A synthetic variety is developed in cross pollinated crops such as maize. For development of synthetic variety, first good general combining inbreds are identified. All inbreds with good general combining ability are inter-mated in all possible combinations and hybrid seed is obtained. The hybrid seed of all combinations is mixed together in equal quantity to constitute a synthetic variety. Thus initially, a synthetic variety consists of several heterozygotes. This is maintained by open pollination which allows some degree of selfing. Hence in later generations, a synthetic variety consists of several homo and heterozygotes. Thus synthetic variety has a heterogeneous population and wide adaptation. The yield level of synthetic variety is always higher than open pollinated variety but lower than the yield of single and double cross hybrids.

Composite Variety

It is developed in cross pollinated species. It is developed by mixing the seed of various open pollinated genotypes which are similar in height, maturity, seed color, seed size, *etc.* Thus composite variety has a heterozygous and heterogeneous population. It consists of several homozygotes and heterozygotes. Such varieties are highly adaptable to environmental fluctuations.

Multiline Variety

It is developed in self-pollinated crops. A multiline variety is a mixture of several isogenic lines, related lines or unrelated lines. Each component line has different gene for disease resistance. Such varieties are homozygous but heterogeneous and have very good buffering capacity due to broad genetic base.

Mutant Variety

A mutant variety is developed through induced mutagenesis. Mutant varieties are generally developed in self-pollinated and asexually propagated species and rarely in cross pollinated species. A mutant variety differs from parent variety in one or two characters. Such varieties are developed to induce disease resistance, earliness, dwarfness and achieve improvement in quality.

Transgenic Variety

A variety which is developed by the technique of genetic engineering is called transgenic variety. Such varieties are developed to solve specific problems that are difficult to be solved by conventional methods of plant breeding. Transgenic varieties with herbicide resistance, insect resistance and improved quality have been developed in several crop plants. Such varieties can be developed in all the three groups of crops *viz.* self-pollinated, cross-pollinated and asexually propagated.

Pedigree, Bulk and Backcross Bred Varieties

Pedigree and bulk bred varieties are developed in self-pollinated species. Such varieties are developed from the progeny of a single best homozygote. Hence varieties developed by pedigree and bulk methods are homozygous and homogeneous and have narrow genetic base. Backcross method is used both in self and cross pollinated species for improvement of oligogenic character such as disease resistance. Backcross bred variety is similar to parent variety except for the gene which has been transferred into parent variety.

Hybrid Variety

The F_1 population that is used for commercial cultivation is called hybrid variety. Hybrid varieties are developed both in cross and self-pollinated species. However, hybrid varieties are more common in the cross pollinated species than in self-pollinated species. Hybrids are of different types, *viz.*, inter-varietal hybrids (single, three way and double cross hybrids) and interspecific hybrids. Hybrids are heterozygous but homogeneous. For hybrid varieties, fresh seed has to be produced every year.

All the varieties discussed above are used for commercial cultivation. These varieties can be divided into different groups on the basis of several criteria. The main criteria which are used for classification of varieties include: (1) country of origin, (2) area of adaptation, (3) genetic constitution, (4) use of seed, (5) evolution techniques, *etc.* (Table 1.1).

Institutions Involved in Release of Variety

In India, new crop cultivars are developed by the following four important organizations:

1. Crop Research Institutes/centres of the Indian Council of Agricultural Research located in different states.
2. State Agricultural Universities located in different states.
3. All India coordinated projects of different crops, and
4. Private seed companies (national and multinational).

New varieties of a crop are developed after thorough testing in All India coordinated project over several locations for 2-3 years. The main objective of coordinated project of each crop is to release productive cultivars.

TABLE 1.1: Classification of Crop Varieties

Sl.No.	*Basis of Classification*	*Types of Variety*	*Brief Description*
1	Country of Origin	Exotic Variety	Introduced from other country.
		Indigenous Variety	Developed within the country.
2	Area of Adaptation	National Variety	Cultivated in several States.
		Local Variety	Cultivated in small area.
3	Use of Seed	Straight Variety	Seed can be used for several years.
		Hybrid Variety	Fresh seed is used every year.
4	Evolution Technique	Simple Variety	Developed by conventional method.
		Transgenic Variety	Developed by genetic engineering.
5	Pollination	Self Pollinated	Maintained by self pollination.
		Cross Pollinated	Maintained by cross pollination.

Variety Release Procedure

In India, the release of new crop varieties consists of four major steps, *viz.* (1) development of new strains, (2) evaluation of performance, (3) identification of superior strains, and (4) release and notification. These are briefly discussed below:

1. Development of New Strains

The new strains are developed by ICAR Crop Research Institutes and State Agricultural Universities for specific purposes. Various breeding methods are used for development of new strains in self and cross pollinated species.

2. Evaluation of Performance

The performance of newly developed strains is evaluated in AICCIP. ICAR Institutes, SAUs and private registered seed companies enter their improved strains/ hybrids in the AICCIP of respective crop for multi-location testing. The new strains are tested at multi-locations under the coordinated project for a minimum period of three years/seasons. The new variety is first tested for yield under the initial varietal trials (IVT); the promising ones are promoted to the advanced varietal trials (AVT) in the second year. The strains that give good performance in AVT for two years are selected.

3. Identification of Superior Strains

The strains which show good yield performance in AVT are identified as superior strains and are considered for release in the workshop meetings. The newer agro and plant protection techniques required to obtain potential yield of new strains are also worked out by that time. The workshop after considering the new promising varieties recommends them to replace existing varieties.

4. Release and Notification

The proposal for release of new varieties is put up in a prescribed proforma to variety release committee. There are two types of variety release committees, *viz.* State Variety Release Committee (SVRC) and Central Variety Release Committee (CVRC). In case of state variety release committee, Director of Agriculture for field crops and Director of horticulture for vegetable and horticulture crops is the chairman. In central variety release committee, Deputy Director General (Crop Science) of ICAR is the Chairman. The release proposal of varieties recommended for All India release is put up before CVRC, while for those recommended for release in a particular state is placed before the SVRC of respective state. These committees consist of scientists, and representatives of seed producing organizations (NSC, SSC and SSCA) and other related government agencies. After release, the variety is notified. Seed production can be taken up only after notification of new varieties. The notification is done by the Government of India.

TABLE 1.2: Location of Coordinated Projects of some Field Crops

Sl.No.	*Crop*	*Location*	*Institute/University*
1	Wheat and Barley	Karnal	Wheat Institute
2	Rice	Cuttack	Rice Institute
3	Maize	Ludhiana, Punjab	Maize Research Institute
4	Sorghum	Hyderabad	Indian Institute of Millets Research
5	Pearl millet	Mandore	Agricultural Research Station.
6	Cotton	Coimbatore	CICR, Regional Station.

Sl.No.	Crop	Location	Institute/University
7	Sugarcane	Lucknow	Sugarcane Institute
8	Potato	Shimla	Potato Institute
9	Chick pea and Pigeon pea	Kanpur	Pulses Institute
10	Soybean	Indore	Soybean Directorate
11	Groundnut	Junagarh	Groundnut Directorate
12	Castor	Hyderabad	Oilseeds Research Institute
13	Linseed	Kanpur	CSA University
14	Sunflower	Bangalore	USA, Campus
15	Rapeseed and Mustard	Bharatpur	Mustard Research Directoraate

Some Abbreviations

SVRC — State Variety Release Committee

CVRC — Central Variety Release Committee

ICAR — Indian Council of Agricultural Research

SAUs — State Agricultural Universities

AICCIP— All India Coordinated Crop Improvement Projects

IVT — Initial Varietal Trial

AVT — Advanced Varietal Trial

CVT — Coordinated Varietal Trial

Summary

A variety refers to a genotype or group of genotypes which have been released for commercial cultivation either by State Variety Release Committee (SVRC) or Central Variety Release Committee (CVRC) and notified by the government of India. A variety is also termed as cultivar (Cultivated variety). A variety has three important characteristics *viz.* (1) distinctiveness (2) uniformity, and (3) stability. Besides distinctness, uniformity and stability, the new variety should be superior to the previously released varieties of a crop in one or more characteristics.

In plant breeding various types of cultivars are developed and released for commercial cultivation. The development of new variety depends mainly on two factors, *viz.* (1) mode of reproduction of the crop species, and (2) method of breeding used for development of a variety. Some varieties are developed by selection without hybridization such as pure line variety, mass selected variety and clonal variety. Development of some varieties involves hybridization and selection. This includes varieties that are developed by pedigree, bulk and backcross methods including multiline cultivars. After release, all these varieties remain under cultivation for several years. The seed of some varieties has to be produced every year afresh such as hybrid varieties and triploids in seed propagated species (*i.e.* watermelon). Some varieties are developed by special breeding techniques such as mutant and transgenic varieties. Different types of varieties which are developed in plant breeding include pure line variety, mass selected variety, clonal variety, synthetic variety, composite variety, multi-line variety, mutant variety, transgenic variety and hybrid variety. These have been explained.

In India, new crop cultivars are developed by four main organizations, *viz.* ICAR research institutes, State Agricultural Universities, all India coordinated projects of different crops, and Private seed companies (national and multinational).

In India, the release of new crop varieties consists of four major steps, *viz.* (1) development of new strains, (2) evaluation of performance, (3) identification of superior strains, and (4) release and notification. These have been discussed.

Glossary

Clonal Variety: In asexually propagated crops [such as potato, sugarcane, sweet potato, *etc.*], a variety developed from the progeny of a single best clone.

Composite Variety: In cross pollinated species a variety developed by mixing the seed of various open pollinated genotypes which are similar in height, maturity, seed color, seed size, *etc.*

Hybrid Variety: The F_1 population that is used for commercial cultivation.

Multiline Variety: In self-pollinated crops a variety which is a mixture of several isogenic lines, related lines or unrelated lines.

Mutant Variety: A mutant variety is developed through induced mutagenesis.

Pure line Variety: In self-pollinated crops a variety developed from a single best pure line progeny.

Synthetic Variety: In cross pollinated species a variety developed using good general combining inbreds.

Transgenic Variety: A variety which is developed by the technique of genetic engineering.

Variety: A genotype released for commercial cultivation either by State Variety Release Committee (SVRC) or Central Variety Release Committee (CVRC) and notified by the government of India.

QUESTIONS

1. **Define variety. What are the norms that are used for release of a new variety?**
2. **Name the varieties which are developed without involving hybridization. Discuss briefly their main features.**
3. **List the varieties which are developed through hybridization and selection. Explain briefly their genetic constitution.**
4. **List the varieties which differ in one or few characters from the parent material. Explain briefly their methods of development and main features.**

5. **Define pure-line, multiline and synthetic varieties. Describe briefly their genetic constitution and methods of breeding.**

6. **Give a brief classification of crop varieties with suitable examples.**

7. **List the institutions that are associated with release of new crop varieties. Explain the role of coordinated projects in the release of new crop varieties.**

8. **Describe briefly various steps that are involved in the release of new crop varieties.**

9. **Write short notes on the following:**

   (*a*) Pure line variety (*b*) Multiline variety

   (*c*) Mass selected variety (*d*) Mutant variety

10. **Differentiate between the following:**

    (*a*) Pure line and clonal variety (*b*) Synthetic and composite variety

    (*c*) Mutant and segregant (*d*) Inter-varietal hybrids and interspecific hybrids.

11. **Define the following:**

    (*a*) Hybrid variety (*b*) Transgenic variety

    (*c*) Exotic variety (*d*) National variety

12. **Explain briefly the genetic constitution of following varieties :**

    (*a*) Mass selected variety (*b*) Synthetic variety

    (*c*) Pure line variety (*d*) Clonal variety

* * * *

Chapter 2

Categories of Improved Seed

Introduction

The seed of a release and popular variety produced by scientific method is referred to as improved seed or quality seed. Variety refers to a genotype which has been released for commercial cultivation either by State Variety Release Committee or Central Variety Release Committee. Improved seed plays an important role in maximizing production and productivity of field crops. It results in (i) better germination, (ii) vigorous seedling growth, (iii) higher crop stand, (iv) better quality of produce, and (v) ultimately in higher crop yield.

Seed Quality

Seed quality is measured in terms of genetic purity, physical purity, germination and seed health. The seeds that meet required standard of genetic purity, good health and physiological purity (viability and vigor) and other attributes are referred to as quality seed.

Characters of Quality Seed

There are certain characteristics of quality seeds such as released variety, genetic purity, seed health, free from inert matter and mixture, undamaged seeds, seed viability, *etc.* These are explained as follows:

1. **Released Variety:** Seed should be of released and notified varieties and their duration should be suitable for the agro-climatic conditions and cropping system of the locality.
2. **Genetic Purity:** Quality seeds are genetically pure (true to type) with high sowing quality, good yield potential, evenness in growth pattern, maturity and should meet the purpose of evaluation.

3. **Good Seed Health:** The seed should be free from seed borne diseases and physiological disorders due to deficiency of plant nutrients or bleaching due to adverse weather.
4. **Proper Size:** The seed should be large, plump, bold uniform in size, shape, color, texture and have proper test weight.
5. **Free from Inert Matter:** The seed should be clean and free from inert matter such as dirt, trash a soil particles, stones, *etc.*
6. **Free from Mixture:** It should be free from mechanical admixture, objectionable weed seed, other crop seeds and insects.
7. **Unbroken Seeds:** It should be whole, not broken, half rotten, half-filled affected with damp.
8. **Fresh Seeds:** The seed should be as fresh as possible or of the proper age. It should not be too old.
9. **Adequate Moisture:** The seed should contain required amount of moisture.
10. **Seed Viability:** The quality seeds are physiologicall by good in terms of germination, vigor, viability and seed health.

Benefits of Quality Seed

There are several advantages of quality seeds such as higher yield, good quality of produce, good germination, crop uniformity, responsive to fertilizers, better adaptability, long life of the variety, aesthetic value and high price. These are briefly discussed as follows:

1. **Higher Yield:** Quality seeds are genetically pure (true to type). They are free from admixture, weed seeds and other crop seeds and insects. Hence we get higher yield from quality seeds. Moreover the prediction of yield of good quality seeds is possible.
2. **High Quality:** We get high quality produce from crop planted with good quality seed. Moreover, the finished product of such seed is also of high quality.
3. **Good Germination**: Quality seeds are vigorous and have high germination due to good seed viability. This results in higher crop stand. It results in fast and uniform emergence of seedling. They are physiologically good in terms of germination, vigor, viability and sound health.
4. **Crop Uniformity:** The crop planted with quality seed is uniform in plant population and maturity. They are free from objectionable weeds, admixture, weed seeds,other crop seeds and insects.
5. **Responsive to Fertilizers**: The quality seeds respond well to the application of fertilizers and nutrients.
6. **Adaptability:** Quality seeds can adopt themselves to extreme climatic condition and cropping system of the location.

7. **Prolong Life of Variety:** Good quality seeds prolong life of a variety. In other words, deterioration of the variety will be slow.
8. **Aesthetic Value:** Crops raised with quality seed are aesthetically pleasing.
9. **High Price:** The produce of quality seed has high produce value which has better marketability.

Classes of Improved Seed

There are five classes or categories of improved seed, *viz.*, nucleus seed, breeder seed, foundation seed, registered seed and certified seed. A brief description of all these in relation to production, purity, certification and use is presented as follows:

1. Nucleus Seed

Nucleus seed is the initial seed of a released and notified variety which is always limited in quantity. It is the initial handful of seeds obtained from selected individual plants of a particular variety. It is also known as basic seed or primordial seed. It is available with the evolving institute with all the pass port information. Main features of nucleus seed are given below:

Production: Nucleus seed is produced at the research farms of originating research institute or agricultural university. Its production is carried under the supersivion of original plant breeder or sponsored plant breeder.

Purity: It is genetically and physically cent per cent pure.

Certification: Certification is not required for the production of nucleus seed.

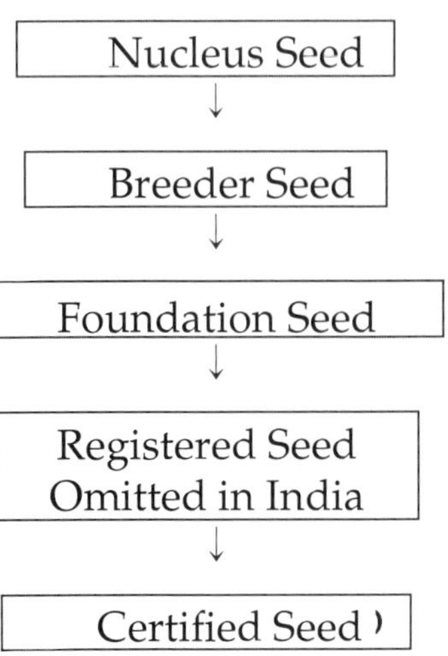

Figure 2.1: Classes or Categories of Improved Seed.

Use: Nucleus seed is used for the production of breeder seed. It is not meant for general distribution.

2. Breeder Seed

Breeder seed is the progeny of either nucleus seed or breeder seed. Main features of nucleus seed are given below:

Production: Breeder seed is produced at the research farms of originating research institute or agricultural university. It is produced under the direct supervision of original plant breeder or sponsored plant breeder.

Purity: It is genetically and physically cent per cent pure. The genetic purity is maintained by isolation and proper rouging. The isolation distance varies from crop to crop.

Certification: Certification is not required for the production of breeder seed. The BSP plot is inspected by monitoring team which consists of a nominee from the Nodal Officer for breeder seed production, Deputy Director of Seed Certification, Area Manager, National Seed Corporation and the Producing breeder. The team inspects the crop during flowering and maturity stages. The breeder seed after passing the field and seed standards should be bagged with breeder seed tag containing all details about crop, variety, lot number, date of test, physical and genetic purity, germination percentage. Grow out test to determine the genetic purity of the seed lot should be conducted for all the breeder seed produced by the breeder. The team assures about genetic purity, isolation distance and general crop condition.

Use: Breeder seed is used for the production of foundation seed. It is not meant for general distribution.

Color of Tag: The seed bags of breeder seed have tag of yellow color.

3. Foundation Seed

Foundation seed is the progeny of breeder. This class of seed should be produced in the area of adaptation of the concern variety. Main features of nucleus seed are given below:

Production: Foundation seed is produced by the National Seeds Corporation under the strict supervision of research scientists and expert from NSC. The seed production is carried out at the research farms of ICAR institutes, agricultural university, Government farms and also on cultivators' fields. Proper isolation distance is adopted for the production of foundation seed which varies from crop to crop.

Types: The foundation seed is of two types, *viz.*, foundation seed stage I and stage II. The former is produced from the breeder seed and latter from the foundation seed.

Purity: Foundation seed is genetically cent per cent pure. However, physical purity of 98 per cent is permissible. The remaining 2 per cent include inert matter.

Certification: This class of seed also requires certification which is undertaken by State Seed Certification Agency.

Use: Foundation seed is used to produce certified seed. It is not meant for general distribution.

Color of Tag: The seed bags of foundation seed have tag of white color.

4. Registered Seed

Registered seed is the progeny of either foundation seed or registered seed. In India, registered seed is generally omitted and certified seed is produced directly from foundation seed. Main features of registered seed are given below:

Production: It is produced at the farms of registered seed growers according to technical advice and supervision of National Seeds Corporation. Proper isolation distance is adopted for the production of foundation seed which varies from crop to crop.

Purity: Registered seed is genetically cent per cent pure. However, physical purity of 98 per cent is permissible. The remaining 2 per cent include inert matter.

Certification: Certification is required for the production of registered seed which is undertaken by State Seed Certification Agency.

Use: Registered seed is used to produce certified seed. It is also not meant for general distribution.

5. Certified Seed

Certified seed is the progeny of either foundation seed or certified seed. Certified seed may be the progeny of certified seed provided this reproduction does not exceed three generations beyond foundation seed stage-I. Main features of certified seed are given below:

Production: Anybody willing to come forward to produce certified seed can produce certified seed. At present State Seeds Corporations, National Seeds Corporation, State Farm Corporation of India, State Departments of Agriculture, Private Companies, Cooperatives and individual farmers are producing certified seed. Proper isolation distance is adopted for the production of certified seed which varies from crop to crop.

Types: The certified seed is of two types, *viz.*, certified seed stage I and stage II. The former is produced from foundation seed and latter from certified seed stage I.

Purity: Certified seed has genetic purity of 100 per cent and physical purity of 98 per cent. Other crop seeds and weed seeds should not be more than prescribed standards which vary from crop to crop.

Certification: Certification is required for the production of certified seed which is undertaken by State Seed Certification Agency.

Use: Certified seed is available to the farmers for commercial crop production. If the quantity of certified seed stage I is not sufficient for general distribution, certified seed stage II is produced from the certified seed stage I. Certified seed cannot be used to produce certified seed again without the approval of the state seed certification agency, which can approve production only under extreme conditions.

Color of Tag: The seed bags of certified seed have tag of blue color.

Maintenance of Nucleus and Breeder Seed

Methods of maintaining nucleus seed/breeders seed in self-pollinated species can be conveniently divided into two groups, *viz.* (i) maintenance of newly released varieties and (ii) maintenance of established varieties. These are discussed as under:

1. Maintenance of Nucleus Seed of Pre-released or Newly Released Varieties

The procedure outlined by Harrington (1952) for the maintenance of nucleus seed of pre-released or newly released varieties is described below:

(i) **Selection of Varieties**: Already released or newly released varieties are selected for nucleus seed production. These varieties are used as base material for nucleus seed production. Nucleus seed is used for further seed production chain. Not more than fifteen new varieties of a crop at one station should be selected for nucleus seed production in one year.

(ii) **Examination of Varieties**: Minimum 200 single plants true to the type of each variety are selected from breeder seed plot, threshed separately and the seed is examined in pile (lots) on the table. Un acceptable piles such as diseased and off types are discarded. The progeny of each of 200 selected single plants is grown separately and evaluated for major characters of the variety. The seed of true breeding progeny is bulked to constitute nucleus seed. Progeny containing off type plants are rejected.

(iii) **Selection of Field:** Each nucleus seed should be grown on clean fertile land at an experiment station in the region or in area in which this new variety could be grown, in the event of its release. The land must not have had a crop of the same kind in the previous year.

(iv) **Inspection of Nucleus Seed Plots**: The nucleus seed plot should be critically examined during the entire crop season from the seedling stage until maturity. Differences in the habit of early plant growth, leaf color, rate of growth, time of heading, plant height, head characteristics and diseases reactions should be critically examined. If a progeny differs distinctly from the average in the pre-heading stages of growth, it should be removed before heading.

(v) **Harvesting and Threshing**: Each remaining plot, of which there should be at least 180 plant progeny out of the original 200 should be harvested individually with a sickle and tied in a bundle. The total bundles of each nucleus should be labeled and stored until the current years yield tests for trials are obtained. The nucleus bundles of any new variety should be discarded, if it is found unworthy of being continued.

After threshing, the seed should be cleaned in a fanning mill or by hand methods. Then the grains from each nucleus plot are placed in a pile on the seed table. The 180 or more piles of seed of one nucleus must be examined for uniformity of seed appearance, and any pile, which appears to be off type be discarded. All the remaining piles of the seed should be bulked together in one lot. This should be treated with fungicide and insecticide, bagged, labeled and stored as "Breeder's Stock Seed" for use in the next year.

2. Maintenance of Breeder's Seed of already Established Varieties

The following steps are involved in the maintenance of breeder's seed.

(i) **Selection of Field:** Breeder's stock seed from the nucleus should be sown on the clean, fertile land, which did not grow a crop of the same kind in the previous year. The space required for seeding the breeder's stock is about 1.2 ha in the case of wheat and as much as 3 ha in the case of transplanted rice.

(ii) **Isolation:** The field should properly be isolated.

(iii) **Agronomic Practices:** The best farm procedures should be used in the sowing, raising and harvesting of breeder's stock.

(iv) **Production:** It should be produced at the experiment station in the area in which the new variety has been bred.

(v) **Sowing:** The breeder seed is always limited in quantity. Hence seeding should be done in such a way as to make the best use of the limited amount of seed available The row spacing should be sufficient to permit examination of plants in rows for possible mixture or off types.

(vi) **Roguing:** In the breeder seed plot, there may a few plants which are off types or different from the variety under multiplication. Such plants should be removed. There should be very few plants to rogue out if the previous year's nucleus or breeder seed was well protected from natural crossing, careful roguing was done and there were no impurities during cleaning, *etc.* The rouging should be done before flowering as well as after flowering.

(vii) **Harvesting and Threshing**: When the breeders seed is harvested and threshed, it should be ensured that the equipment used must be thoroughly cleaned and free from seeds of any other varieties. This cleanliness should be observed during cleaning and bagging also. The seed thus produced would be about 99.9 per cent pure as to variety. This breeder seed is ready now for increase of foundation seed. A portion of this breeders' seed should be retained by the breeders for further production of breeder seed of the variety.

3. Maintenance of Breeders' Seed of Established Varieties

The breeder's seed of established varieties could be maintained satisfactorily by any one of the following methods.

(i) **By Raising the Crop in Isolation**. The breeder's seed of local varieties can be maintained by growing them in isolated plots and by very rigorous roguing during various stages of crop growth, where the various plant characters are observable. The method of handling the breeder seed crop is the same as described earlier for breeder's seed of newly released varieties.

(ii) **By Bulk Selection**. The genetic purity of established varieties can be satisfactorily maintained by bulk selection. In this method 2,000 to 2,500 plants typical of the variety are selected, harvested and threshed separately. The seeds from each plant are examined and any Pile (heap) which shows any obvious off-types, or otherwise appears dissimilar, are discarded. The remaining piles of seed are bulked to constitute the breeder's seed. The other practices of handling breeder seed production remains the same as explained above.

TABLE 2-1: Differences between Breeder Seed and Certified Seed

Sl.No.	*Particulars*	*Breeder Seed*	*Certified Seed*
1	Progeny of	Nucleus or breeder seed	Foundation or certified seed
2	Produced by	Original/sponsored breeder	State Seed Corporations
3	Produced at	Research or Govt. farms	Farmers' fields
4	Genetic purity	100 per cent	99.9 per cent
5	Physical purity	100 per cent	98 per cent
6	Certification	Not required	By SSCA
7	Used for	Production of foundation seed	Commercial crop production
8	Color of tag	Yellow	Blue
9	Stage II used	Sometimes	Often

Carry-over Seed

The breeder must carry-over enough seed of the variety to safeguard against the losses caused by complete crop failure during the foundation seed multiplication phase. In addition, the breeder should further safeguard the variety by storing a portion of the original seed under ideal conditions.

Summary

The seed of a release and popular variety produced by scientific method is referred to as improved seed or quality seed. Variety refers to a genotype which has been released for commercial cultivation either by State Variety Release Committee or Central Variety Release Committee. Improved seed plays an important role in maximizing production and productivity of field crops. It results in (i) better germination, (ii) vigorous seedling growth, (iii) higher crop stand, (iv) better quality of produce, and (v) ultimately in higher crop yield.

Seed quality is measured in terms of genetic purity, physical purity, germination and seed health. The seeds that meet required standard of genetic purity, good health and physiological purity (viability and vigor) and other attributes are referred to as quality seed. Characteristics and benefits of good quality seeds have been discussed.

There are certain characteristics of quality seeds such as released variety, genetic purity, seed health, free from inert matter and mixture, undamaged seeds, seed viability, *etc.* These points have been explained.

There are several advantages of quality seeds such as higher yield, good quality of produce, good germination, crop uniformity, responsive to fertilizers, better adaptability, long life of the variety, aesthetic value and high price. These are briefly discussed.

There are five classes or categories of improved seed, *viz.*, nucleus seed, breeder seed, foundation seed, registered seed and certified seed. A brief description of all these in relation to production, purity, certification and use has been given. Methods of maintenance of nucleus and breeder seeds in self-pollinated species have been discussed.

Glossary

Breeder seed: The progeny of either nucleus seed or breeder seed produced under the direct supervision of original or sponsoring plant breeder.

Certified seed: The progeny of either foundation seed or certified seed.

Foundation seed: The progeny of breeder seed produced by NSC or SSC.

Improved Seed: The seed of a release and popular variety produced by scientific method; also called quality seed.

Nucleus seed: The initial seed of a released and notified variety which is always limited in quantity.

Registered seed: The progeny of either foundation seed or registered seed. This category is omitted in India.

Variety: A genotype which has been released for commercial cultivation either by State Variety Release Committee or Central Variety Release Committee.

QUESTIONS

1. **What is Nucleus? Describe the procedure of nucleus seed production.**
2. **Define breeder seed. Describe various aspects of breeder seed production.**
3. **What is Foundation seed? Describe the procedure of foundation seed production.**
4. **Define registered. Describe the procedure of registered seed production.**
5. **What is certified seed? Describe the procedure of certified seed production.**
6. **Describe Nucleus seed on the basis (i) production, (ii) purity, (iii) certification, and (iv) uses.**
7. **Describe Breeder seed on the basis (i) production, (ii) purity, (iii) certification, and (iv) uses.**

8. **Describe Foundation seed on the basis (i) production, (ii) purity, (iii) certification, and (iv) uses.**

9. **Describe certified seed on the basis (i) production, (ii) purity, (iii) certification, and (iv) uses.**

10. **Define the following terms:**

    (i) Nucleus seed

    (ii) Breeder seed

    (iii) Registered seed

    (iv) Foundation seed

11. **Explain various characteristics of quality seed.**

12. **Describe briefly benefits of quality seed.**

* * * *

Chapter 3

Indian Seeds Act 2004

Introduction

The Indian Seeds Bill 2004 hereinafter called Indian Seeds Act 2004 was approved by the government of India in 2004 and came into force with effect from January 2005. The main objective of the Seeds Bill, 2004 is to ensure availability of quality seeds to farmers. It replaces the Seeds Act, 1966.

Main Features

The main features of Indian Seed Act 2004 in relation to coverage, seed category, license, registration, criteria for protection, period of protection, certification, compensation, farmers' privilege, penalty, exclusion, research exemption and constitution of central seed committee are briefly presented below:

1. **Coverage:** This Act has wider coverage than seeds act of 1966. It includes agriculture, horticulture, forestry, plantation crops, medicinal and aromatic plants. The seeds of all these crops are covered by the new act.
2. **Seed Category:** This act permits sale of certified seed only. The name of variety, physical purity and germination percentage have to be indicated on the seed container or bag.
3. **License:** The seed dealers, sellers and growers should have license which is compulsory.
4. **Registration:** Registration for all kinds and varieties of seed is compulsory. No producer can grow seeds unless he is registered. It is compulsory for all cultivated varieties of field crops, vegetable crops, fruit crops, medicinal and aromatic plants, forest species and plantation crops. Every seed producer and dealer, and horticulture nursery has to be registered with the state government.
5. **Criteria for Registration**: The released and notified varieties can be registered for seed multiplication. There are four criteria for protection

of varieties under PPV and FR Act. These are novelty, distinctiveness, uniformity and stability.

6. **Period of Registration**: The Indian Seeds Act 2004 has adopted UPOV Act 1978. According to this Act, the period of protection is 15 years for annual and biennial crops and 18 years for perennial plants such as trees and vines.
7. **Certification:** The Act permits self certification of seeds by accredited agencies and allows the central government to recognize certification by foreign seed certification agencies. Seed producers are permitted to self certify the performance of their seed under certain conditions. The seed companies need to provide the results of multiplication trials before registration.
8. **Compensation:** The disputes about the quality of seeds have to be settled in the consumer court. Any loss in the production due to poor seed quality is claimed in the consumer court.
9. **Farmers' Privilege:** The Seeds Act 2004 allows farmers' privilege. The Bill does not restrict the farmer's right to use or sell his farm seeds and planting material, provided he does not sell them under a brand name. All seeds and planting material sold by farmers will have to conform to the minimum standards of germination, physical purity and genetic purity applicable to registered seeds.
10. **Penalty**: Any person who contravenes any provisions of the Act or imports, sells or stocks seeds deemed to be misbranded or not registered, can be punishable by a fine between Rs 5,000 and Rs 25,000. The penalty for giving false information is a prison term up to six months and/or a fine up to Rs 50,000. In case of companies, the person(s) in charge of conduct of the business of the company will be held accountable.
11. **Exclusion**: Those techniques which are harmful to animals and the environment such as terminator technology and traitor technology are excluded and not allowed by the seed act.
12. **Research Exemption:** This act allows use of protected material by breeders for development of new varieties.
13. **Central seed Committee**: In the Seed Act 2004, the central seed committee consists of Chairman [Secretary DARE], Agricultural Commissioner, DDG horticulture [ICAR], Joint Secretary Seeds[DAC], representatives from DBT, Ministry of Environment and Forest, Secretary Agriculture [from five States], Director [SSCA] from one State and MD-SSC from one State. In addition, two representatives each of farmers and seed industry are nominated.

Comparison of Seeds Act 1966 and Seeds Act 2004

There are some similarities and some dis-similarities between two seeds acts which are presented in Table 3.1.

TABLE 3.1: Comparison of Seed Act 1966 and Seed Act 2004

Sl.No.	*Particulars*	*Seeds Act 1966*	*Seeds Act 2004*
A	**Similarities**		
1	Category of seed permitted for sale	Certified Seed	Certified Seed
2	License	Compulsory	Compulsory
3	Certification	Compulsory	Compulsory
4	Farmers' exemptions	Available	Available
5	Purity and germination standards	Adopted	Adopted
6	Compensation of loss by use of certified seed	Through consumer court	Through consumer court
7.	Requirements for seed multiplication	Released and notified varieties	All seeds for sale must be registered
B	**Dissimilarities**		
1.	Crop covered	Agriculture and horticulture	Agriculture, horticulture, forestry, plantation crops, medicinal and aromatic plants
2	Registration of transgenic varieties	No provision	Special provision
3	Registration with PPV and FR authority	Not required	Required
4	Period of protection	Not defined	Defined
5	Penalty for violation of Act	Rs 100-1000/and 6 months imprisonment	Rs 500-50,000/and six months imprisonment
6	Self Certification	Not permitted	Permitted
7	Seed multiplication of land Races	Permitted	Not permitted
8	Representatives in central seed committee	From all States	From five States only
9	Involvement of Private Sector	No	Yes

Comparison of Seeds Act 2004 and PPV & FR Act 2001

There are some similarities and some dis-similarities between Seed Act 2004 and Protection of Plant Variety and Farmers' Rights Act 2001 which are presented in Table 3.2.

Merits of Seed Act 2004

1. **Wider Coverage**: This act has wider coverage [agriculture, horticulture, forestry, plantation crops, medicinal and aromatic plants] than Seed Act of 1966.
2. **Easy Certification**: This act has made the process of certification very simple, means allowed self certification. But it should not be misused.

TABLE 3.2: Comparison of Seeds Act 2004 and Protection of Plant Varieties and Farmers' Rights Act 2001

Sl.No.	*Particulars*	*Seeds Act 2004*	*PPV and FR Act 2001*
A	**Similarities**		
1	Registration	Compulsory	Compulsory
2	Period of registration	15 years for annual and biennial crops and 18 years for perennials	15 years for annual and biennial crops, 18years for perennials
3.	Renewal of registration	Permitted for similar period	Permitted for similar period
4	Farmers Rights	Allowed	Allowed
5	Plant breeders' Rights	Available	Available
6	Disclosure of variety's performance	Compulsory	Compulsory
7	Sale of branded seeds by farmers	Not permitted	Not permitted
B	**Dissimilarities**		
1	Record Maintained	National Register of Seeds	National register of plant varieties
2	Compensation	Through consumer court	Through PPV and FR Authority
3	Disclosure of variety's parentage	Not required	Required
4	Variety's performance to be declared by	Seed producer, distributor or vendor	Breeder of the variety
5	Penalty for violation of the Act	Rs 5000-50,000/	Rs 50,000-20 Lakh
6	Imprisonment for violation of Act	Up to six months	From 2-3 years

3. **Ban Harmful Technologies**: The act did not permit registration of harmful technologies such as terminator technology and traitor technology.
4. **Permits Farmers' Rights**: It permits farmers' rights. The farmers need not to register their varieties.
5. **Ensures Quality Seed**: This act ensures availability of good quality seeds to farmers by imposing heavy penalties on the sale of spurious seeds.
6. **Registration for long Duration**: This act provides registration for 15 years for annual and biennial crops and 18 years for perennial crops which is sufficiently long period. Moreover, the registration is renewable for another term of similar duration.
7. **Central Seed Committee:** In the Seed Act 2004, the constitution of central seed committee is better than that of Seed Act 1966.
8. **Promotes Seed Industry:** The new seed act will lead to fast development of seed industry due to involvement of representatives of multinational seed companies or industries in the central seed committee.

9. **Increase in Productivity:** The new act will help in increasing crop productivity due to availability of better quality seeds.
10. The present seed replacement rate is 20-25 per cent. The seed act will enhance seed replacement rate of various crops.
11. The seed act will boost the export of seeds and encourage import of useful germplasm for use in crop Improvement programs.
12. It will encourage investment in seed research and development.

Demerits of Seed Act 2004

Although farmers are exempt from registering their seed varieties, the seeds have to conform to standards prescribed for commercial seeds. Farmers may find it difficult to adhere to the standards required of commercially sold seeds, because the seed saved and exchanged by farmers constitute above 80 per cent of the seeds planted.

1. It is not clear whether the compensation would include the value of the crop or only the cost of the seed.
2. It is not clear whether a seed producer may sell seed which is registered by a different producer.
3. The disclosure of parentage is not essential. It may lead to a situation where seeds can be registered without disclosing the parentage or origin of the seed.
4. The Act does not have the provision of benefit sharing unlike the Convention on Biological Diversity and the PPVFR Act.
5. The Seed Act does not provide for a mechanism to trace back a packet of seed to the dealer, processor and producer. This would make it difficult to trace back a defective lot, and rectify any deficiencies in the supply chain.
6. Seed producers would be permitted to self-certify the performance of their seeds under certain conditions. This opens up the possibility of false declaration by seed companies. To prevent this, only government agencies should be allowed to conduct these trials and grant certification.
7. Every horticultural nursery has to be registered with the state government and has to maintain records of layout plan, source of every planting material, *etc.* Such nurseries in the unorganized sector may find it difficult to adhere to these conditions.
8. The Seed Inspector has the power to enter and search as well as break open container or break open doors, without a warrant. This is against the normal procedure.
9. The new seeds act will encourage cultivation of new varieties. Thus old varieties and land races will be replaced by new varieties on vast areas resulting in significant reduction in biodiversity. Reduction in biodiversity will invite danger of uniformity and lead to narrow genetic base and narrow adaptation.

10. It will encourage multinational seed companies which may lead to monopoly of such companies. Seed companies may hike seed prices resulting in exploitation of farmers.

Summary

The Indian Seeds Bill 2004 hereinafter called Indian Seeds Act 2004 was approved by the Government of India in 2004 and came into force with effect from January 2005. The main objective of the Seeds Bill, 2004 is to ensure availability of quality seeds to farmers. It replaces the Seeds Act, 1966. The main features of Indian Seed Act 2004 in relation to coverage, seed category, license, registration, criteria for protection, period of protection, certification, compensation, farmers' privilege, penalty, exclusion, research exemption and constitution of central seed committee are presented.

The comparison of Seeds Act 1966 and Seeds Act 2004 a and Protection of Plant Varieties and Farmers' Rights Act 2001 is presented in tabular form. The merits and demerits of Seed Act 2004 have been highlighted.

Glossary

Container: A box, bottle, casket, tin, barrel, case, receptacle, sack, bag, wrapper or other thing in which any article or thing is placed or packed.

Dealer: A person who carries on the business of buying and selling, exporting, or importing seed, and includes an agent.

Essentially derived variety: A variety derived from the initial variety and having most of the characteristics of initial variety but clearly distinguishable from initial variety.

Export: Taking out of India by land, sea or air.

Extant variety: A variety available in India such as farmers' variety, variety of common knowledge or any other variety of public domain.

Farmer: Any person who cultivates crops either by cultivating the land himself or through any other person but does not include any individual, company, trader or dealer who engages in the procurement and sale of seeds on a commercial scale.

Horticultural Nursery: Any place where horticulture plants are,produced or propagated and sold for transplantation.

Import: Bringing into India by land, sea or air;

Kind: One or more related species or sub-species of crop plants each individually or collectively known by one common name such as cabbage, maize, paddy and wheat.

Misbranded: A seed with false label or label of another variety.

Producer: A person, group of persons, firm or organization who grow or organize the production of seeds.

Seed processing: The process by which seeds and planting materials are dried, threshed, shelled, ginned or delinted (in cotton), cleaned, graded or treated.

Seed: Any type of living embryo or propagule capable of regeneration and giving rise to a plant of agriculture which is true to such type.

Spurious seed: Any seed which is not genuine or true to type.

Transgenic variety: Seed or planting material synthesized or developed by modifying or altering the genetic composition by means of genetic engineering.

Variety: A plant grouping within a single botanical *taxon* of the lowest known rank.

QUESTIONS

1. **Explain in brief main features of Indian Seeds Act 2004.**
2. **What are merits and demerits of Indian Seeds Act 2004?**
3. **Give a comparison of Indian Seeds Act 1966 and Seed Act 2004 in a tabular form.**
4. **Compare Indian Seeds Act 2004 and Protection of Plant Varieties and Farmers' Rights Act 2001.**
5. **Describe briefly main features of Indian Seeds Act 1966.**
6. **Discuss briefly the advantages and limitations of Indian Seeds Act 1966.**
7. **Describe the constitution of Central Seed Committee of (i) Indian Seeds Act 1966 and (ii) Indian Seeds Act 2004.**
8. **What do the following terms signify?**
   (i) Extant Variety (ii) Transgenic variety
   (iii) Spurious Seed (iv) Farmers' Rights
9. **Write short notes on the following:**
   (i) Farmers' Rights (ii) Farmers' Compensation
   (iii) Orthodox seeds (iv) Recalcitrant seeds
10. **Describe briefly Indian Seeds Act 2004 on the basis of following**
    (*a*) Seed certification (*b*) Registration
    (*c*) Licensing (*d*) Penalties
11. **Describe briefly the Structure of Central Seed Committee of Indian Seed Act 2004.**
12. **What are the similarities and dissimilarities between Seeds Act 1966 and Seeds Act 2004?**
13. **Describe briefly the differences between Indian Seeds Act 2004 and Protection of Plant varieties and farmers' Rights Act 2001.**

14. Write Short Notes on the following:

(i) Indian Seeds Act 1966

(ii) Indian Seeds Act 2004

(iii) PPV and FR Act 2001

(iv) Biodiversity Act 2002

* * * *

Chapter 4

Protection of Plant Varieties and Farmers' Rights Act 2001

Introduction

The Protection of Plant Varieties and Farmers' Rights Act 2001 was approved by the Government of India in 2004 but came into force with effect from January, 2005. The main objective of this act is to encourage development of new plant varieties and promote development of seed industry in India.

Main Features

The main features of PPV and FR Act 2001 in relation to purpose, legislative structure, headquarters, registration, period of registration, farmers' rights, plant breeders' rights, exclusion, compensation and penalties are briefly presented below.

1. **Purpose:** There are four main objectives of enforcing the Protection of Plant varieties and Farmers' Rights Act 2001 as given below:
   - (i) To establish an effective system for protection of plant varieties, farmers' rights and plant breeders' rights.
   - (ii) To recognize and protect farmers' rights for their contribution in conserving, improving and making available plant genetic resources for development of new plant varieties.
   - (iii) To stimulate investment for research and development both by public and private sector for development of new plant varieties.
   - (iv) To promote the growth of seed industry in the country to ensure availability of high quality seeds to the farmers.
2. **Legislative Structure:** The Authority consists of a chair person called as authority and 15 persons as members. The members of the Authority include (i) Agricultural Commissioner-GOI, (ii) DDG Crop Sciences-

ICAR, (iii) Joint Secretary Seeds-GOI, (iv) Horticulture Commissioner-GOI, (v) Director, NBPGR; (vi) one member each [not below the rank of Joint Secretary] from DBT, (vii) Ministry of Environment and forests-GOI, (viii) Ministry of Law Justice and Company Affairs- GOI; (ix) one representative each [nominated by central government] from farmers, (x) tribal organization, (xi) seed industry, (xii) an agricultural university, (xiii) and women organization; and (xiv) two nominated representative of state government.

3. **Headquarters:** The headquarters of Protection of Plant Varieties and Farmers' Rights Authority is located in Pusa Campus [National Agricultural Science Complex], New Delhi.
4. **Plant Material Protected**: Various types of crop cultivars can be protected under PPV and FR Act. The following breeding material can be protected.
   (i) All new varieties of self pollinated species such pure line and multi line varieties.
   (ii) New varieties of cross pollinated species such as open pollinated variety, synthetics and composites.
   (iii) New varieties of vegetatively/clonally propagated species.
   (iv) Single, three-way and double cross hybrids.
   (v) Inbred parental lines of hybrid varieties.
   (vi) Trees, fruits, ornamental plants and vines, *etc.*
5. **Registration:** The PPV and FR Act Authority establishes a Plant Varieties Registry, which maintains a National Register of Plant Varieties. It specifies details under which a variety may be registered such as a complete passport data of the parental lines from which a variety has been derived have to be furnished by the breeder. A new variety will be registered under the act if it conforms to the criteria of novelty, distinctiveness, uniformity and stability.
   (i) **Novelty:** It refers to newness of a variety. A variety which has not been grown for more than one year prior the application for registration can be protected.
   (ii) **Distinctiveness:** The variety should be clearly distinguishable fro previously available varieties at least for one characteristics.
   (iii) **Uniformity**: It refers to same type of population of a variety. In cross pollinated species, the range of variation should be within the tolerable limit.
   (iv) **Stability:** It means that the variety should give stable performance after repeated reproduction or propagation. Moreover, a variety should have a designation.

   For registration of a variety, an application has to be filed with the Registrar, PPV and FR Act. No fee is charged if a variety is registered by a farmer.

6. **Period of Protection**: The PPV and FR Authority has accepted UPOV Act 1978. According to this Act, the period of protection is 15 years for annual and biennial crops and 18 years for perennial plants such as trees and vines. The registration is renewable after its expiry for a similar term of duration.
7. **Farmers' Rights:** A farmer is entitled to save, use, sow, re-sow, exchange, share or sell his farm produce including seed of a variety protected under the Act in the same manner before this Act came into force. He cannot sell branded seed of a variety protected under the Act.
8. **Plant Breeders' Rights:** A certificate of registration for a variety issued by this act shall confer an exclusive right on the breeder or his successor, his agent or licensee, to produce, sell, market, distribute, import or export seed of his variety. The breeder has right to authorize any other person for production and marketing of his variety.
9. **Exclusion:** Those techniques which are harmful to animals and the environment such as terminator technology and traitor technology are excluded and not allowed by the Protection of Plant Varieties and farmers' Rights Act 2001.
10. **Compensation:** If a breeder of a propagating material of a variety registered under the Act sells his product to a farmer, he has to disclose the expected performance under given conditions. If the propagating material fails to perform, the farmer can claim compensation in the prescribed manner before the Protection of Plant Varieties and Farmers' Rights Authority.
11. **Penalty**: Penalty for applying false denomination to a variety is imprisonment up to two years and/or a fine between Rs 50,000 and Rs five lakh. Penalty for falsely representing a variety as registered is imprisonment up to three years and/or a fine between Rs one lakh and Rs five lakh or both. Penalty for subsequent offence is imprisonment up to three years and/or a fine between Rupees two lakh and Rupees 20 lakh.

Merits of PPV and FR Act 2001

It provides an effective system for protection of plant varieties, farmers' rights and plant breeders' rights.

1. It recognizes the contribution of farmers in conserving, improving and making available plant genetic resources for development of new plant varieties.
2. It provides rights to farmers to save, use, sow, re-sow, exchange, share or sell his farm produce including seed of a variety protected by PPV and FR Act.
3. It confers an exclusive right on the breeder or his successor, his agent or licensee, to produce, sell, market, distribute, import or export seed of his variety. Thus it gives recognition to breeders for their contribution and encourage them to breed still better varieties.

4. It encourages investment in plant breeding research and development both by public and private sectors for development of superior plant varieties. It will lead to fast development of seed industry and improvement in seed quality due to competitions among breeders.
5. It promotes the growth of seed industry in the country to ensure availability of high quality seeds to the farmers.
6. It involves representatives of farmers, women and seed industry in its legislative structure who will look to the interest of these sectors.
7. The Act has provision for compensation. If the seed gives poor performance than the declared by the breeder, farmer can claim compensation in the prescribed manner before the Protection of Plant Varieties and Farmers' Rights Authority.
8. The Act has provision for rewarding farmers for their role in conserving plant genetic resources. Such expenditure can be met out from the gene fund available with the PPV and FR Act Authority.
9. The act has provision of benefit sharing. The farmers can get share from the benefits arising from cultivation of farmers varieties and knowledge.
10. The act has provision to impose penalties to curb unhealthy practices.

Demerits of PPV and FR Act 2001

1. The PPV and FR Act will promote cultivation of new high yielding varieties which are highly uniform. This will result in reduction in genetic diversity leading to narrow genetic base and danger of uniformity.
2. This will encourage monopoly of some breeders or seed companies for genetic material with special traits
3. The holders of plant breeders' rights will produce less seed than market demand and hike price to earn more profit. The high price cannot be afforded by small and marginal farmers.Thus all farmers cannot take advantage of new varieties.
4. It will inhibit free exchange of seed material and encourage unlawful practice.

Summary

The Protection of Plant Varieties and Farmers' Rights Act 2001 was approved by the Government of India in 2004 but came into force with effect from January 2005. The main objective of this act is to encourage development of new plant varieties and promote development of seed industry in India. The main features of PPV and FR Act 2001 in relation to purpose, legislative structure, headquarters, plant material protected, registration, period of registration, farmers' rights, plant breeders' rights, exclusion, compensation and penalties have been discussed.

There are four basic requirements for protection of a variety under PPV and FR Act, *viz.* (i) novelty, (ii) distinctiveness, (iii) uniformity, and (iv) stability. Varieties, hybrids and inbred lines of hybrids are protected under this act.

A comparison of Indian Seeds Act 2004 and Protection of Plant varieties and farmers' Rights Act 2001 has been given in tabular form. Merits and demerits of the act have been highlighted.

Glossary

Candidate Variety: A variety that has to be protected.

Distinctiveness: The variety should be clearly distinguishable fro previously available varieties at least for one characteristic.

Example variety: A variety that is used for comparison of a particular character.

Extant variety: All released and notified varieties which have not been protected.

Farmers' Variety: A variety that has been developed by a farmer and used for commercial cultivation for several years.

Novelty: Newness of a variety. A variety which should have not been grown for more than one year prior the application for its registration.

Reference variety: All released and notified extant varieties that are under seed production chain

Stability: The same performance of a variety after repeated reproduction or propagation.

Uniformity: The same type of population of a variety.

QUESTIONS

1. **Describe briefly main features of PPV and FR Act 2001**

2. **Discuss briefly merits and demerits of PPV and FR Act 2001**

3. **Compare Indian Seed Act 2004 and Protection of Plant Variety and Farmers' Act 2001.**

4. **Explain briefly legislative structure of PPV and FR Act 2001.**

5. **Write short notes on the following:**

   (i) Plant Breeders Rights (ii) Transgenic variety

   (iii) Farmers Rights (iv) Candidate variety

6. **Describe briefly PPV and FR Act 2001 on the basis of following**

   (i) Period of Protection (ii) Criteria for Registration

   (iii) Compensation (iv) Penalties

7. **Describe briefly the requirements for registration of a variety under PPV and FR Act 2001.**

8. **What do the following terms signify:**

   (i) Extant variety (ii) Reference variety

   (iii) Example variety (iv) Farmers' variety

9. Write short notes on the following:

(i) Novelty
(ii) Distinctiveness
(iii) Uniformity
(iv) Stability

10. Give full form of the following:

(i) PBR
(ii) PVPA
(iii) DUS
(iv) UPOV

* * * *

Chapter 5

DUS Testing

Introduction

Evaluation of plant varieties in terms of distinctiveness, uniformity and stability is referred to as DUS testing. The testing of DUS characters is useful in the following four main ways.

1. For Identification of plant varieties.
2. For Registration and protection of plant varieties under Protection of Plant Varieties and Farmers Rights Act 2001.
3. For varietal information system and classification of varieties in different groups, and
4. For Creating data base for plant genetic resources.

Government of India has passed an Act entitled "Protection of Plant Varieties and Farmers' Rights Act, 2001". The main mandates of the Act are:

1. Registration of plant varieties.
2. Characterization, cataloguing and documentation of registered varieties.
3. Ensuring that seeds of all registered varieties are made available to farmers.
4. Maintenance of National Register of Plant varieties.

Farmers' Rights

Provision of Farmers' Rights in the Act is unique to the Indian system of Plant Variety Protection. It provides rights to the farmers to:

(i) Register their varieties.

(ii) Share benefit for use of biodiversity conserved by farming community.

(iii) Save, use, sow, re-sow, exchange, share or sell farm produce including seed of registered variety (not branded).

(iv) Claim compensation for under performance from the promised level.

(v) Seek their consent when farmer variety is used to develop an essentially derived variety (EDV).

Breeders Rights

Breeders' rights allow researchers to use protected variety for the purpose of research and breeding new varieties and for commercializing them with out securing prior permission from the PBR holder of the initial variety.

Compulsory Denomination

Any protected variety must be given a single and distinct denomination in accordance with the regulations. Any person who markets the produce of this protected variety must use the same denomination even if the period of protection is expired which is again a valuable part of Breeders' rights.

Types of Varieties

In connection with plant variety protection (PVP) Act, various terms such as extant variety, candidate variety, reference variety, example variety and farmers variety are frequently used. Hence, knowledge of these terms is essential. These are defined below:

- ☆ **Extant Variety**: All released, notified and unprotected varieties.
- ☆ **Candidate variety:** A variety to be registered under Plant Variety Protection Act is referred to as candidate variety.
- ☆ **Reference Variety:** All released and notified extant varieties of common knowledge which are in seed production chain.
- ☆ **Example Variety:** A variety that is used for comparison for a particular character is called example variety.
- ☆ **Farmers' Variety**: A variety that has been developed by a farmer and used for commercial cultivation for several years is called farmers variety.

Requirements for Protection

The PPV and FR Act, 2001 provides that protection shall only be granted after examination of the variety for Distinctness, Uniformity and Stability and Novelty. The first three requirements were as per UPOV Act, 1978 and are known as DUS. The fourth was included in UPOV Act, 1991 and known as NDUS.

1. **Distinctiveness**: The new variety must be clearly distinguishable by one or more characters from any other variety whose existence is a matter of common knowledge at the time when the protection is applied for.
2. **Uniformity**: The variety should be sufficiently uniform in its relevant characteristics. Relevant characteristics include all those used as a basis for distinctiveness or included in the variety description established at the date of grant of protection of that variety.

3. **Stability:** The variety is deemed to be stable if its relevant characteristics remain unchanged after repeated propagation.
4. **Novelty:** It refers to newness of a variety. The variety should be new one and should not have been commercially cultivated for more than one year before granting protection under PVP Act.

Period of Protection

The period of protection varies with plant species. For field crops, the minimum period of protection is 15 years, whereas for forest trees, fruit trees, ornamental trees, shrubs and vines it is 18 years.

How to Claim PBR?

Registration of any new plant variety under the PPV FR is voluntary. It consists of following steps.

1. An application is to be submitted in the prescribed form with varietal description bringing out Distinctness, Uniformity, Stability and Novelty.
2. Complete passport data of parental lines, their geographical origin is to be furnished.
3. An affidavit declaring that the variety does not contain terminator gene sequence and it was lawfully acquired.
4. Specified fee for registration and conducting DUS testing is to be paid.
5. Prescribed quantity of seed of the variety and parental lines (for hybrids) is to be deposited.

DUS Testing in Cotton

The details of DUS testing in cotton are briefly discussed as follows.

Material Required

The minimum quantity of seeds to be supplied along with application is 2 Kg for varieties, hybrids and parental line in cotton which has to be provided in one submission. The seeds should meet the minimum requirement of germination (> 80 per cent), moisture content (< 10 per cent) and physical purity (98 per cent for varieties, hybrids and their parents) and highest genetic purity. The seed should not have undergone any treatment unless the authority allows it.

Conduct of Test

The minimum duration of tests should normally be two independent similar growing seasons at two locations. The test conditions should ensure normal growth of the plant. The plot design is as follows:

No. of rows: 12	Row length: 6 m
Row to row distance: 90 cm	Plant to plant distance: 60 cm
No. of replications: 3	

Methods of Observations

The characteristics described in the table of characteristics should be used for the testing of varieties, inbred lines and hybrids. The observations should be made on 40 plants which should be divided among 4 replications (10 plants per replication). For uniformity, testing is done by visual assessment of a group of plants in single observation. The number of aberrant plants should not exceed 8 in 150. All leaf characteristics should be observed on 4th leaf from the top.

DUS Testing Centers

In cotton, four DUS testing centers, *viz.* Nagpur, Coimbatore, Dharwad and Hisar have been identified. The Nagpur and Hisar centers will deal with testing of *G. hirsutum* and *G. arboreum* species. Coimbatore centre will test *G. hirsutum* and *G. barbadense* species and Dharwad centre will test varieties of *G. hirsutum* and *G. herbaceum.*

Characters for DUS Testing

As per National DUS Test Guidelines, in cotton, 41 characters have been decided for DUS testing. These traits are related to seedling (1), plant (3), stem (2), leaf (9), bract (2), flower (6), boll (8), fibre (7) and seed (3) and include both oligogenic as well as polygenic traits.

The characteristics are again classified into three groups, *viz.* (i) Grouping characters, and (ii) Essential characters and (iii) Additional characters.

Grouping Characters

These are highly heritable characters which are generally not found to vary within variety under any condition and hence suitable for grouping of varieties into different classes. Grouping characters include leaf shape, petal colour, boll shape and fibre length (Table 5.1).

TABLE 5.1: Characters for Grouping in Cotton

1.	Leaf shape	Palmate (normal), semi-digitate (semi-okra), digitate (okra) and laceolate (super okra)
2.	Petal colour	White, cream, yellow, pink, red and dark red
3.	Boll shape	Round, oval and elliptic
4.	Fibre length	Very short, short, medium, long and extra-long

Essential Characters

These are traits which have been showing the most variation among the varieties and are necessarily to be recorded and included in the variety description. In cotton, there are 20 essential characteristics as follows which also includes the grouping characters [Table 5.2].

TABLE 5.2: Essential Characters for DUS Testing in Cotton

1.	Leaf shape	Palmate, Semi–digitate, digitate and super-okra
2.	Leaf size (width at maximum point)	Small, medium and large
3.	Leaf colour	Light green, green, light red and dark red
4.	Leaf : Pubescence	Absent, medium and strong
5.	Leaf nectaries	Absent and present
6.	Bract type	Normal and frego
7.	Petal colour	White, cream, yellow, pink, red and bi-color
8.	Petal spot	Absent and present
9.	Position of stigma	Embedded and exerted
10.	Anther colour	White, cream, yellow and purple
11.	Boll size	Small, medium and large
12.	Boll shape	Round, oval, elliptical
13.	Boll surface	Smooth and pitted
14.	Boll opening	Open, semi-open and close
15.	Fibre length	Very short, short, medium, long and extra-long
16.	Fibre colour	White, cream, brown and green
17.	Fibre strength	Weak, medium and strong
18.	Ginning per cent	Low, medium, high and very high
19.	Seed : Density of fuzz	Naked, semi-fuzzy and fuzzy
20.	Seed: Fuzz colour	White, grey, brown and green

Optional Characters

There are 21 characters that may or may not be recorded as per requirement and are given in the table of characteristics. List of optional characters along with their categories is presented in Table 5.3.

TABLE 5.3: Optional Characters for DUS Testing in Cotton

1.	Hypocotyl : pigmentation	Absent and present
2.	Plant : time of flowering (50 per cent of plants with at-least one open flower)	Early (<45 days), medium (45-60 days) and late (>60 days)
3.	Plant : stem pigmentation	Absent and present
4.	Plant : stem hairiness	Absent, sparse, medium and strong
5.	Leaf : lobe number	One, three, five and seven
6.	Leaf : appearance	Cup and flat
7.	Leaf : gossypol glands	Absent and present
8.	Leaf : petiole pigmentation	Absent and present
9.	Bract : number of serration	Few, medium and many
10.	Flower : Male sterility	Absent and present
11.	Flower : filament pigmentation	Absent and present
12.	Boll : bearing habit	Solitary and cluster

13.	Boll : color	Green and red
14.	Boll : prominence of tip	Blunt and pointed
15.	Boll : weight of seed cotton/boll	Small (<3.0 g), medium (3.1-5.0 g.) and large (>5.0 g)
16.	Plant : growth habit	Determinate and indeterminate
17.	Plant : height	Very short (<61 cm), short (61-90 cm), medium (91-120 cm), tall (121-150 cm) and very tall (>150 cm)
18.	Seed : size (100 seed wt.)	Very small (<5.1 g), small (5.1-7.0 g), medium (7.1-9.0 g), bold (9.1 – 11.0 g) and very bold (> 11 g)
19.	Fibre : fineness (micronaire value)	Very fine (<3.0), fine (3.0-3.9), medium (4.0-4.9), coarse (5.0-5.9) and very coarse (> 5.9)
20.	Fibre : uniformity	Poor (<40), average (40-45) and good (>45)
21.	Fibre : maturity (per cent)	Poor (<70), average (70-80) and good (>80)

There are three more additional characters which are not yet included in the table of characteristics but could be observed as an additional character if sufficient variability could not be obtained from the above characteristics. These are vein/sepal pigmentation, bract pigmentation and petiole pigmentation.

Summary

DUS testing refers to evaluation of varieties in terms of distinctiveness, uniformity and stability. DUS testing is useful for identification of varieties, registration of varieties under PPV and FR Act 2001, varietal information system and creating data base for plant genetic resources. There are four characters which are used for grouping. Such characters include leaf shape, petal color, boll shape and fibre length. These are highly heritable characters that are not influenced by environmental effects. For DUS testing, characters are classified in to two groups, *viz.* (i) essential characters [20 characters] and (ii) optional characters [21 characters]. Essential characters have to be recorded, whereas optional characters may or may not be recorded.

In cotton, there are four DUS testing centres, *viz.*, CICR, Nagpur, CICR (RS), Coimbatore, UAS Dharwad and HAU, Hisar. Nagpur and Hisar centres will carry out testing of *G. hirsutum* and *G. arboreum;* Coimbatore for *G. hirsutum* and *G. barbadense;* and Dharwad for *G. hirsutum* and *G. herbaceum*. The test has to be conducted at least at two locations for two normal seasons. Each test should be based on minimum sample of 150 plants which have to be divided into four replications.

Glossary

Candidate variety: A variety to be registered under Plant Variety Protection Act.

Distinctiveness: The new variety must be clearly distinguishable by one or more characters from any other variety whose existence is a matter of common knowledge at the time when the protection is applied for.

DUS Testing: Evaluation of plant varieties in terms of distinctiveness, uniformity and stability.

Example Variety: A variety that is used for comparison for a particular character.

Extant Variety: All released, notified and unprotected varieties.

Farmers' Rights: Legal rights provided to the farmers to save, use, sow, replant, exchange, share or sell his farm produce including seed of a variety protected under Plant Variety Protection Act.

Farmers' Variety: A variety that has been developed by a farmer and used for commercial cultivation for several years.

Novelty: It refers to newness of a variety. The variety should be new one and should not have been commercially cultivated for more than one year before granting protection under PVP Act.

Plant Breeders' Rights: Legal rights granted to the plant breeder for a new plant variety. It provides rights for commercial seed production, marketing, export and import, authorization and to prevent infringement.

Reference Variety: All released and notified extant varieties of common knowledge which are in seed production chain.

Stability: The variety is deemed to be stable if its relevant characteristics remain unchanged after repeated propagation.

Uniformity: The variety should be sufficiently uniform in its relevant characteristics. Relevant characteristics include all those used as a basis for distinctiveness or included in the variety description established at the date of grant of protection of that variety.

QUESTIONS

1. **Define DUS testing. What are the criteria for DUS testing?**
2. **Explain briefly Farmers' rights and Plant Breeders' rights.**
3. **Define the following terms:**
   (i) Candidate Variety (ii) Example Variety
   (iii) Extant Variety (iv) Farmers' Variety
   (v) Reference Variety.
4. **Describe the period of protection for various plant species as per UPOV Act 1978.**
5. **Explain various characters used for DUS testing in cotton.**
6. **Write short notes on the following.**
   (i) Farmers Rights (ii) Plant Breeders Rights

7. **Define the following terms:**

(i) Distictiveness

(ii) Uniformity

(iii) Stability

(iv) Novelty

* * * *

Chapter 6

Seed Certification

Introduction

Seed certification is a legally sanctioned system for the quality control of seed multiplication and production. Certificate is granted by a seed certification agency after a seed dealer files an application. The Seed Certification Agency is the legally sanctioned system for quality control of seed in a country. It implements quality control schemes, and ensures good quality of seeds with minimum purity and germination standards. In general, seed certification is a process designed to maintain and make available to the general public continuous supply of high quality seeds and propagating materials of released and notified varieties of crop plants.

Brief History

The concept of seed certification originated in the beginning of the twentieth century. The Swedish workers were first to initiate the process of field evaluation of the seed crops. In the beginning, agronomists and plant breeders visited the fields of progressive farmers who took the seeds of new varieties from them. This initiated the process of field inspection and found to be very helpful in keeping varieties pure in the seed production chain. Subsequent important events related to seed certification are presented in table 6.1.

TABLE 6.1: Important Events Related to Seed Certification

Sl.No.	Year	Events
1	1919	Scientists from USA and Canada met in Chicago, Illinois and formed an International Crop Improvement Association (ICIA),
2	1969	Changed the name of International Crop Improvement Association to Association of Official Seed Certifying Agencies (AOSCA) paving the way for modern day seed certification.
3	1963	In India the field evaluation of the seed crop and its certification started with the establishment of National Seeds Corporation.

Sl.No.	Year	Events
4	1966	A legal status was given to seed certification with the enactment of first Indian Seed Act.
5	1968	Seed Rules were formulated.
6	1970	The first official State Seed Certifications Agency was established in Maharashtra.
7	1974	Karnataka was the first State to establish the Seed Certification Agency as an autonomous body.
8	1976	National Seed Project was launched to establish State Seeds Corporations.
9	2004	The new Seed Act was introduced to replace Seed Act of 1966.
10	2005	The Seed Act 2004 came in force in January, 2005.

Now, there are 21 States Seed Certification Agencies in the country which have been established under the Seed Act, 1966. In great majority of the countries in the World, including India, seed certification is voluntary and labelling is compulsory.

Purpose of Seed Certification

The purpose of seed certification is to maintain and make available to the public, through certification, high quality seeds and propagating materials of notified kind and varieties so grown and distributed as to ensure genetic identity and genetic purity. Seed certification is also designed to achieve prescribed standards.

The main objective of the Seed Certification is to ensure the acceptable standards of seed viability, vigour, purity and seed health. A well organized seed certification should help in accomplishing the following three primary objectives.

(i) Systematic increase of superior varieties;

(ii) Identification of new varieties and their rapid increase under appropriate and generally accepted names.

(iii) Provision for continuous supply of comparable material by careful maintenance.

Certified Seeds

Certified seed is produced from foundation, registered or certified seeds. This is so known because it is certified by a seed certification agency *i.e.* State Seed Certification Agency. The certified seed is annually produced by progressive farmers according to standard seed production practices. To be certified, the seed must meet the prescribed requirements regarding purity and quality. These standards vary from one crop to another.

Requirements for Certified Seed

Seed has to meet certain rigid requirements before it can be certified for distribution. The important requirements for seed certification are listed as follows.

1. An Improved Variety/hybrid.
2. Genetic purity

3. Physical purity
4. Germination percentage
5. Freedom from weeds
6. Freedom from diseases and pests
7. Optimum moisture contents

1. Improved Variety

The first and foremost requirement is that the seed must be of an improved variety released by either the Central or a State Variety Release Committee for general cultivation and notified by the Ministry of Agriculture, Government of India; this is essential for the seed to be certified.

2. Genetic Purity

Genetic purity means the absence of seeds of other varieties of the same crop species as well as of other crop species. Genetic purity ensures that the seed is of the variety under certification, and that there is no mixture from other varieties or other

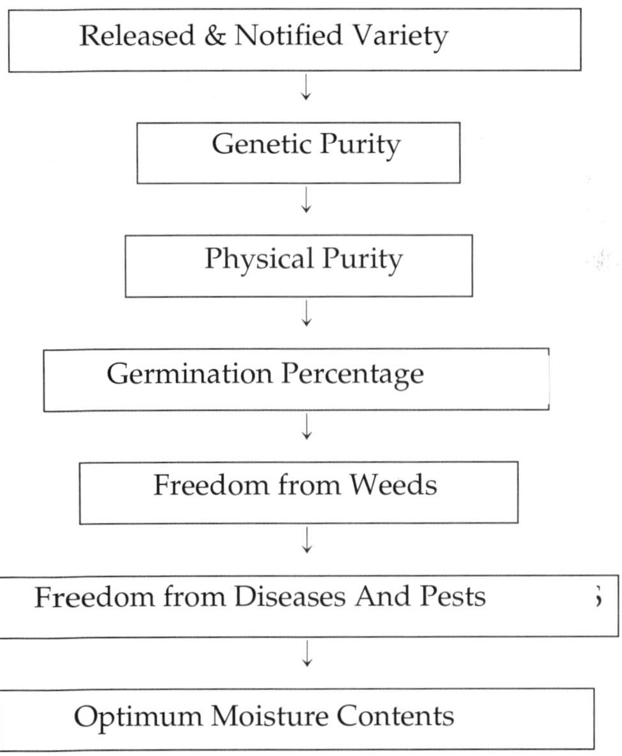

FIGURE 6.1: Rcquirements for Seed Certification.

crops. The standard of genetic purity is very high; the amount of contamination permitted ranges from 0-0.1 per cent. In most of the cases, contamination by seeds of other crop species is permitted to a small degree (up to 0.1 per cent), but contamination by seeds of other varieties of the same crop is generally not permitted.

3. Physical Purity

Physical purity implies freedom of seed from inert matter and defective seed. Inert matter consists of nonliving material such as sand, pebbles, soil particles, straw, *etc.* Defective seeds are those seeds that are broken, diseased, insect infested, shrivelled and unfit for germination. A broken seed larger than half of the normal seed is not considered defective provided the embryo is not damaged. The total amount of permissible contamination by inert matter and defective seed ranges from 1.0 per cent in the case of maize (Zea mays) to 5 per cent in the case of carrot.

4. Germination

A high percentage of germination is necessary to obtain a good crop stand with the minimum amount of seed. The percentage of germination required for certification is high in a crop like maize (90 per cent), moderate in several others *e.g.*, jowar (80 per cent), wheat (85 per cent), rice (80 per cent), *etc.* and relatively low in many others, *e.g.* cauliflower (65 per cent), lady finger (65 per cent), carrot (60 per cent) and chillies (60 per cent). Thus the minimum standard of germination of seed certification varies considerably from one crop to the other.

5. Freedom from Weed Seeds

Freedom from weed seeds is necessary to prevent weeds from spreading through seed and to reduce losses caused by weeds. The maximum amount of weed seeds permitted is very low; it varies from zero per cent in crops like maize (*Z. mays*) and tomato (*L. esculentum*) to 0.2 per cent in cauliflower (*B. oleracea*), onion (*A. cepa*), carrot (*D. carota*) *etc.* For some crop species, some weeds are classified as objectionable or noxious weeds. In such crops, the certified seed is required to be free from seeds of the noxious weed species.

6. Freedom from Diseases

The certified seed must be free from seed-borne diseases. If the seed is contaminated with pathogens, it is likely to lead to an epidemic and a total loss of the crop. The presence of seed-borne pathogens is prevented in two ways. Firstly by effective control of diseases in the standing crops, and secondly by treating seeds with insecticides and fungicides.

7. Optimum Moisture Content

The seeds must be dried to an optimum moisture level for efficient processing and safe storage. The level of optimum moisture varies from one crop species to another. It varies from 13 per cent in case of rice to 7 per cent in the case of cauliflower and watermelon (*Citrulla vulgaris*). A moisture content higher than the recommended level may have following effects.

1. Loss in seed viability
2. Increase in the growth of microorganisms, particularly moulds,
3. Increase in attacks by storage insect pests,
4. Higher chances of damage to seeds by fumigants, and
5. Adverse effects on processing because damp seeds tend to stick together.

Seed Certification Agencies

Seeds Act, 1966 provides for the establishment of Seed Certification Agencies in each State. The organizational set up of the Certification Agency includes Board of Directors, technical and other staff for operating the programme. Seed Certification Agency may have its own seed testing laboratory or it may get its seed samples tested through seed testing laboratories. Seed Certification Agency should function on the following broad principles:

1. Seed Certification agency should be an autonomous body.
2. Seed Certification Agency should not involve itself in the production and marketing of seeds.
3. The Seed Certification Standards and procedures adapted by seed certification agency should be uniform, throughout the country.
4. Seed Certification Agency should have close linkage with the technical and other related institutions.
5. Its long-term objective should be to operate on no-profit no-loss basis.
6. Adequate staff trained in seed certification should be maintained by the Certification Agency.
7. It should have provision for creating adequate facilities for ensuring timely and thorough inspections.
8. It should serve the interests of seed producers and farmers/users.

Seed Certification Measures

Seed certification measures include, verification of seed source, field inspections, sample inspection, bulk inspection, control plot test and grow out test. All these are discussed as follows:

1. **Verification of Source Seed:** Source seed verification is the first step in Seed Certification Programme. Unless the seed is from approved source and of designated class certification agency will not accept the seed field for certification. Seed of only those varieties which are released and notified is eligible for certification. It ensures the use of high quality true to type seed for sowing of seed crops. The individual intending to produce seed under certification shall submit to the Certification Agency, one or more relevant evidence such as certification tags, seals, labels, seed containers, purchase records, sale records, *etc.*, as may be demanded by the Certification Agency during submission of the application, its scrutiny and/or during first inspection of the seed crop, in order to confirm if the

seed used for raising the crop has been obtained from the source approved by certification Agency.

2. **Field Inspection:** The main purpose of field inspections is to examine five things, *viz.* (i) isolation distance, (ii) off types, (iii) objectionable weeds, (iv) disease and insect incidence, and (v) general crop condition. Certified seed is produced by registered seed growers. A producer means a person/organisation who grows or distributes certified seed as per procedures and standards of the certification. Field inspections are made by the inspectors of State Seed Certification Agency to examine the suitability of crop for certification. The number of field inspection varies from 2 to 4 depending upon the crop species. Field inspections are done as follows:
   (i) When only two inspections are made, one is made during flowering and another before harvesting.
   (ii) In case of three inspections, one is done before flowering, second at the time of flowering and third before harvesting.
   (iii) In case of four inspections, one is done before flowering, second at the time of flowering and third and fourth between flowering and harvesting.
3. **Sample Inspection:** It refers **to** assessment of the planting value of the seeds by laboratory tests. Certification agency draws representative samples from the seeds produced under certification programme and subjects them to germination and other purity tests required for conforming to varietal purity.
4. **Bulk Inspection:** Sometimes, bulk inspection is carried to check homogeneity of the bulk seed produced as compared with the standard sample. This gives an idea about the genuineness of lot and sample.
5. **Control Plot Testing:** Here the samples drawn from the source and final seed produced are grown side by side along with the standard samples of the variety in question. This comparison helps in determining the varietal purity and health of the produced seed with original variety.
6. **Grow-Out Test:** This test is carried out to evaluate the seeds for their genuineness to species or varieties or seed borne infection. Here the samples drawn from the lots are grown in the field along with the standard checks. Growing plants are observed for the varietal purity. Grow-out test helps in the elimination of the sub-standard seed lots.

Phases of Seed Certification

Seed Certification is carried out in six broad phases listed as under:

1. Verification of seed source, class and other requirements of the seed used for raising the seed crop.
2. Receipt and scrutiny of application.
3. Inspection of the seed crop in the field to verify its conformity to the prescribed field standards.

4. Supervision at post-harvest stages including processing and packing.
5. Drawing of samples and arranging for analysis to verify conformity to the seed standards; and
6. Grant of certificate, issue of certification tags, labelling, sealing *etc.*

Minimum Seed Certification Standards

The minimum seed certification standards are the standards required for the certification of seeds by the certification agencies. The minimum seed certification standards can be broadly divided into two groups, *viz.*, general seed certification standards and specific crop standards. These two combined sets of standards constitute the minimum seed certification standards for seed certification.

1. General Seed Certification Standards

The general seed certification standard aims at outlining the general requirements for the production of genetically pure seed. These standards prescribe the procedure for production of certified seed with maximum genetic purity and good quality.

TABLE 6.2: Minimum Field Standards of Some Important Crops

Sl.No.	*Crop*	*Seed Class*	*Isolation*	*Field Inspections*	*Off Types*	*Other Crop Plants*
1	Paddy	FS	3	2	.05	--
		CS	3	2	.20	--
2	Wheat	FS	3	2	.05	.01
		CS	3	2	.20	.05
3	Sorghum (OP Varieties)	FS	200, 400*	3	.05** .10***	--
4	Pulses	FS	10	2	.10	--
5	Oilseeds Groundnut	FS	3	2	.10	--
6	Soybean	FS	3	2	.10	--
		CS	3	2	.50	--
7	Fiber Crops Cotton	CS	30	2	.20	--
8	Vegetable Crops Potato Seed	CS	50	4	.10	--
9	Tomato	FS	50	3	.10	--
		CS	25	3	.20	--
10	Cauliflower	CS	1000	3	.20	--
11	Onion	FS	5	2*	.10	--

FS = Foundation Seed, CS = Certified Seed, OC = Other Crop Plants

2. Specific Crop Standards

Specific crop standard consists of field standards and seed standards. Field standards specify:

(i) The minimum preceding crop requirement to minimize genetic contamination from the disease and volunteer plants.

(ii) The minimum isolation distance to minimize out-crossing.

(iii) The number of seed disease infected plants and stage of crop to ensure verification of genetic purity and other quality factors.

Seed Standards Specify:

(i) The minimum percentage of pure seeds and maximum permissible limits for inert matter, other crop seeds.

(ii) The maximum permissible limits for objectionable weeds, seeds infected by seed borne diseases to ensure goods seed health.

(iii) The maximum permissible limits of seed moisture content for the safe storage of seeds.

Summary

Seed certification is a legally sanctioned system for the quality control of seed multiplication and production. Certificate is granted by a seed certification agency after a seed dealer files an application. The Seed Certification Agency is the legally sanctioned system for quality control of seed in a country. It implements quality control schemes, and ensures good quality of seeds with minimum purity and germination standards.

The concept of seed certification originated in the beginning of the twentieth century. The Swedish workers were first to initiate the process of field evaluation of the seed crops. In the beginning, agronomists and plant breeders visited the fields of progressive farmers who took the seeds of new varieties from them. This initiated the process of field inspection and found to be very helpful in keeping varieties pure in the seed production chain. Subsequent important events related to seed certification are presented.

The purpose of seed certification is to maintain and make available to the public, through certification, high quality seeds and propagating materials of notified kind and varieties so grown and distributed as to ensure genetic identity and genetic purity. Seed certification is also designed to achieve prescribed standards.

Seed has to meet certain rigid requirements before it can be certified for distribution. The important requirements for seed certification are an improved variety/hybrid, genetic purity, Physical purity, germination percentage, freedom from weeds, freedom from diseases and pests and optimum moisture contents. All these have been discussed.

Seed certification measures include, verification of seed source, field inspections, sample inspection, bulk inspection, control plot test and grow out test. All these have been discussed. Minimum seed certification standards have been explained.

Glossary

Certified seed: The seed produced from foundation, registered or certified seeds.

Grow-Out Test: The test which is carried out to evaluate the seeds for their genuineness to species or varieties or seed borne infection.

Seed Certification Agency: A government organization which is authorized for seed certification. It may have its own seed testing laboratory or it may get its seed samples tested through seed testing laboratories.

Seed certification measures: Various measures which are adopted to ensure quality seed production such as verification of seed source, field inspections, sample inspection, bulk inspection, control plot test and grow out test.

Seed certification standards: The minimum standards of isolation distance, genetic purity and germination required for the certification of seeds prescribed by the certification agencies.

Seed Certification: It is a legal system which ensures availability of high quality seed of improved variety to the farmers.

QUESTIONS

1. **Define seed certification and describe its objectives.**
2. **Explain briefly history of seed certification with particular reference to India.**
3. **Describe briefly requirements of seed certification.**
4. **Discuss organizational structure and functions of state seed certification agency.**
5. **What are the measures adopted for seed certification?**
6. **Explain briefly various phases of seed certification.**
7. **Describe briefly minimum seed certification standards of ten crops in India.**
8. **Explain the following**

   (i) General seed certification standards (ii) Specific crop standards
9. **What are the criteria taken in to account during seed certification.**
10. **Describe the adverse effects of high seed moisture contents.**

* * * *

Part II

Maintenance Breeding

Chapter 7

Causes of Varietal Deterioration

Introduction

Production of good quality seed is an important task that requires high technical skills, adequate land area and adequate funding. Proper attention has to be paid to maintain genetic purity during seed production to exploit its potentiality in next generation. In other words, seed production must be carried out under standardized and well organized condition. Varietal deterioration refers to permanent reduction either in genetic or agronomic value of a released and notified crop variety.

Causes of Varietal Deterioration

The genetic purity of a variety or trueness to its type deteriorates due to several factors during the seed production cycles. There are several causes of varietal deterioration which can be divided into two groups, *viz.* genetic causes and non-genetic causes. Genetic causes include natural out-crossing, spontaneous mutations, minor genetic variations and developmental variations. Non genetic causes include mechanical mixtures and effect of pests and diseases. Important factors that are responsible for genetic deterioration of crop varieties are listed below (Kadam, 1942).

A. Genetic causes

1. Natural Out-crossing

In sexually propagated species, natural out-crossing with other varieties of the same crop is an important source of varietal deterioration. The extent of contamination depends upon the magnitude of natural cross-pollination. Natural crossing with undesirable types such as off types and diseased plants leads to varietal deterioration. Natural out-crossing is not a serious problem in self-pollinated crops. However, it may lead to contamination if proper isolation distance is not maintained from other varieties. The natural crossing, however, can be major source of contamination in cross-pollinated crops. The extent of genetic contamination in seed field due to natural crossing depends upon several factors such as (i) the breeding system of the species, (ii) isolation distance, (iii) varietal mass and (iv) cross-pollinating insects.

The isolation of seed crops is the most important factor for avoiding contamination of the cross-pollinated crops. The direction of prevailing wind, number of insects present and their activity, and number of varieties are also important considerations in contamination by natural crossing.

2. Spontaneous Mutations

Spontaneous mutations do continuously arise in crop varieties due to cosmic radiations. Such mutations after crossing with true genotype of a variety lead to deterioration of that variety. The impact of natural mutants on varietal deterioration is much lesser than out-crossing. Periodical rouging is the effective way to prevent varietal deterioration from such mutations.

3. Minor Genetic Variations

Sometimes, residual variability also known as minor genetic variations exist even in those varieties looking uniform and homogenous when released. This variation, leads to deterioration of a variety during later cycles of seed multiplication. Hence, due care has to be taken about residual variability at the time of release of a variety. Moreover, it is necessary to pay proper attention during the maintenance of nucleus and breeder's seed of cross-pollinated species.

4. Developmental Variations

When seed crops are grown under different agro-climatic conditions for several consecutive generations developmental variations may set in as differential growth responses. A particular environment may favour some specific genotypes of a variety. This may lead to shift in the value of a variety in a particular direction. Hence, varieties of crops should be grown in the areas of their natural adaptation to minimize developmental shifts.

B. Non genetic Causes

5. Mechanical Mixtures

Mixing of different varieties of the same crop may take place during sowing with seed drills, ginning, harvesting, threshing, seed processing and seed storage. Mechanical mixture is the most important source of varietal deterioration. Mechanical mixture also occurs through volunteer plants of the same crop in the seed field. The mixing also occurs during harvesting and threshing operations. The threshing equipment is often contaminated with seeds of other varieties.

Similarly, the gunny bags, seed bins and elevators are also often contaminated, leading to the mechanical mixtures of varieties. Proper care during seed sowing, harvesting, threshing, processing and seed storage are necessary to avoid such mechanical contamination.

6. Selected Influence of Pest and Diseases

New races of diseases and biotypes of insects are constantly occurring. Sometimes, an improved crop variety deteriorates due to infection of new race of a disease which was not prevailing at the time of release of such variety. New crop varieties often are susceptible to newer races of pests and diseases. This leads to reduction in the genetic value of a variety. Hence, seed production should be carried out strictly under disease free conditions as far as possible.

7. The Techniques of the Plant Breeder

A variety should be critically evaluated for all important attributes before its release because, even minor segregation in the variety for certain characters can cause significant deterioration of such variety. This failure can be attributed to the variety-testing program. Besides these, variations arising due to recombination and natural chromosome doubling may also take place in varieties during seed production, which can be avoided by periodical selection during maintenance of the seed stock.

Genetic Purity Maintenance

Hartmann and Kester (1968) and Agarwal (1980) described steps to maintain the genetic purity of variety during seed production. The following measures have been suggested to safeguard the genetic purity.

1. **Multiplication in Adapted Area**: The seed production should only be carried out in areas of adaptation of a variety to avoid genetic shifts.
2. **Approved Class of Seeds**: The seed multiplication should only be undertaken for approved class of seed, *viz.*, breeder seed, foundation seed and certified seed in a generation system.
3. **Preceding Crop Requirement**: Inspection and approval of seed plots prior to planting is essential to avoid contamination from volunteers and other varieties.
4. **Isolation:** The separation of seed plot from various sources of contamination such as natural crossing or mechanical mixtures is essential. The seed crop must be isolated from other nearby fields from the same crop or any contaminating crop as per certification standards. Isolation of seed production of different varieties is also necessary to avoid mechanical contamination.

5. **Roguing**: Removal of off-types [having phenotypic differences from the variety] from the seed plot is essential to maintain the genetic purity. The number of rouging varies with the crop, purity of the source seed and the stage of the multiplication of the seed crops. Rouging in most of the field crops may be done at - Vegetative, Pre-flowering, Flowering and Maturity stages
6. **Field Inspection**: Qualified and experienced personnel of seed certification agency should inspect seed crops at all appropriate stages of growth and verify seed lots or purity and quality.
7. **GOT**: The genetic purity of the varieties should be tested through grow out test. The GOT should be conducted periodically.

Agronomic Principles

Besides genetic principles, the following agronomic principles should be adopted to maintain good seed quality and obtain higher seed yields.

1. Selection of Variety and Seed

Variety should be high yielder and adapted to agro-climatic condition. Variety should possess other desirable characters like disease resistance, earliness, grain quality, *etc.* The seed used for raising a seed crop should be of known purity, appropriate class and invariably obtained from authorized official agency. If the seed is not treated already, it should be treated with appropriate fungicides/insecticide prior to sowing.

2. Selection of Area and Seed Plot

The seed crops have to be grown only in areas well adapted to the photoperiodic and temperature conditions prevailing. Region of moderate rainfall and humidity are much more suited to seed production than regions of high rainfall and humidity. In general, regions with extreme summer heat and very cold winters should also be avoided for seed production unless particular crops are especially adapted to grow and product under these conditions.

Seed production plot should have good texture and fertility. The fields should be free from volunteer plants, weeds and other crop plants. Soil should be free from soil borne diseases and insect pests. The previous crops should not be same crop. The plot should get adequate isolation distance. The land must be prepared well. Good land preparation helps in improved and uniform germination resulting in good stand establishment.

3. Agronomic Practices

Standard agronomic practices such as sowing time, seed rate, sowing method, weed control and irrigation should be adopted to raise a healthy crop.

Nutrition plays important role in proper development of plant and seeds. Nitrogen is required for good crop growth, phosphorus for proper fruiting and seed development and potash for flowering and seed development.

4. Plant Protection Measures

Recommended plant protection measures should be adopted to minimize losses due to diseases and insects.

5. Harvesting of Seed Crops

Time and method of harvesting are important. The harvesting should be done when the seed is fully matured (Moisture content less than 20 per cent). Early or late harvesting affects seed quality. Every effort should be made to avoid chance of mechanical mixing, mechanical injury during harvesting, threshing, cleaning drying and packing.

6. Drying of Seeds

In order to preserve seed viability and vigor, it necessary to dry seeds to safe moisture content levels as early as possible without heaping wet seed \ pods.

7. Storage of Raw Seeds

After sun drying, the seed should be filled in clean bags. The bags should be dipped in 2 per cent Melathion, dried and cleaned before they are filled. Mark the bag with name of variety and other details. The bags should be stacked on wood pallets but not directly on the floors. The height of the stack should not be more than 3 to 4 m for cereals, 2.5 to 3 m for other crops. The seed stores should be dry, cool and clean and sprayed with Melathion and later fumigate as and when necessary.

Summary

Production of good quality seed is an important task that requires high technical skills, adequate land area and adequate funding. Proper attention has be paid to maintain genetic purity during seed production to exploit its potentiality in next generation. In other words, seed production must be carried out under standardized and well organized condition. Varietal deterioration refers to permanent reduction either in genetic or agronomic value of a released and notified crop variety.

There are several causes of varietal deterioration which can be divided into two groups, *viz.* genetic causes and non-genetic causes. Genetic causes include natural out-crossing, spontaneous mutations, minor genetic variations and developmental variations. Non genetic causes include mechanical mixtures and effect of pests and diseases. All these have been discussed. Genetic and agronomical measures for preventing varietal deterioration have been discussed.

Glossary

Genetic Deterioration: The heritable reduction in the genetic value of a released and notified crop variety.

Isolation Distance: The separation of the field of a variety from that of another variety of the same crop by a minimum prescribed distance to avoid contamination.

Mechanical Mixture: The inadvertent mixing of seed of two varieties during threshing, transport, storage, ginning (in cotton) or sowing with seed-drill.

Outcrossing: Crossing of a variety with the plants of another variety leading to genetic contamination.

Roguing: The process of removal of the off types plants from the field of an improved variety to avoid contamination.

Spontaneous Mutations: Sudden heritable change in the phenotype of an individual.

Varietal deterioration: Permanent reduction either in genetic or agronomic value of a released and notified crop variety.

Volunteer plants: The plants which grow in the field from previous crops.

QUESTIONS

1. **Define varietal deterioration. Describe briefly various causes of varietal deterioration.**
2. **What are genetic causes of varietal deterioration? Describe them in brief.**
3. **Explain briefly non-genetic causes of varietal deterioration.**
4. **Describe genetic measures for preventing varietal deterioration.**
5. **Discuss various agronomical principles that have to be adopted to prevent varietal deterioration.**
6. **Write short note on the following:**
   (i) Genetic causes of varietal deterioration
   (ii) Non genetic causes of varietal deterioration
7. **Explain the following terms:**
   (i) Varietal deterioration (ii) Natural out-crossing

(iii) Spontaneous mutations

8. **Describe the role of natural out-crossing and spontaneous mutations in varietal deterioration.**

9. **Discuss briefly role of mechanical mixtures in varietal deterioration.**

10. **Describe role of isolation distance and rouging in maintaining varietal purity.**

* * * *

Chapter 8

Basics of Maintenance Breeding

Introduction

There are three important areas or activities of plant breeding, *viz.* (i) varietal development, (ii) seed multiplication and (iii) varietal maintenance. These are defined as follows:

1. **Varietal Development:** It makes use of various breeding methods such as selection, hybridization [pedigree, backcross and bulk methods], mutation and genetic engineering to develop new plant varieties.
2. **Seed Multiplication:** It deals with multiplication or production of successive generations of various categories of seeds such as nucleus seed, breeder seed, foundation seed and certified seed. The seed multiplication or increase is regulated by seed certification agencies and breeders.
3. **Varietal Maintenance:** It deals with continuous fresh supply of breeder seed which is used to start another cycle of seed multiplication. It makes use of various selection/maintenance procedures.

Maintenance Breeding

A branch or an area of plant breeding which deals with principles and methods of breeder seed production is referred to as maintenance breeding. In other words, it deals with ways and means of maintaining genetic and physical purity of released and notified varieties and parents of hybrids. Thus maintenance breeding is nothing but the breeding procedure followed to maintain genetic purity of the variety. It is also known as varietal maintenance technology. Main points related to maintenance breeding are presented as follows:

1. Purpose

The main purpose of maintenance breeding is to provide fresh breeder seed of crops every year which is used to start another cycle of seed multiplication.

Maintenance breeding is used to enhance the life span of released and notified varieties and commercial hybrids. Thus, the main objective of maintenance breeding is to make available continuous supply of breeder seed. It has to maintain the original performance of a variety recorded at the time of its release. Thus, it prevents varietal deterioration.

2. Base Material

Nucleus or breeder seed is used as base material for starting maintenance breeding program. When breeder seed and nucleus seeds are in short supply, foundation seed may also be used to start maintenance breeding program.

3. Breeding Methods Used

Various breeding procedures or techniques are used for maintenance breeding. Important breeding procedures used for maintenance breeding in various field crops include positive mass selection, negative mass selection, single plant progeny selection, *etc.* The selection applied for maintenance breeding purposes does not ensure genetic gain in the genetic value of a variety.

TABLE 8.1: Breeding Methods Used for Maintenance Breeding in different Field Crops

Sl.No.	*Name of Crop*	*Breeding Method Used*	*Units Selected*
1	Wheat	Ear to progeny row selection	300 to 500
2	Barley	Ear to progeny row selection	300 to 500
3	Triticale	Ear to progeny row selection	300 to 500
4	Maize	Ear to progeny row selection	100 to 150
5	Pearl millet	Ear to progeny row selection	500 to 1000
6	Sorghum	Panicle to progeny row selection	200 to 500
7	Rice	Panicle to progeny row selection	100 to 500
8	Rapeseed and Mustard	Plant to progeny selection	
	(i) Self-pollinated		400 to 500
	(ii) Cross-pollinated		2000 to 2500
9	Soybean	Plant to progeny selection	400 to 500
10	Groundnut	Plant to progeny selection	100 to 150
11	Sunflower	Plant to progeny selection	500 to 800
12	Sugarcane	Plant to progeny selection	100 Clumps
13	Cotton	Plant to progeny selection	200 to 250
14	Jute	Plant to progeny selection	500 to 600
15	Tobacco	Capsule to progeny selection	1 to 5
16	Potato	Plant to progeny selection	100 to 150
17	Black gram, Green gram	Plant to progeny selection	100 to 150
18	Chickpea, Field pea, Lentil	Plant to progeny selection	100 to 150
19	Pigeon pea	Plant to progeny selection	500 to 1000
20	Castor [CP]	Plant to progeny selection	200 to 500

Sl.No.	*Name of Crop*	*Breeding Method Used*	*Units Selected*
21	Sesame	Plant to progeny selection	400 to 500
22	Linseed	Plant to progeny selection	300 to 400
23	Oats	Panicle to progeny selection	400 to 500
24	Niger	Plant to progeny selection	800 to 1000

4. Main Focus

All the characters of a variety recorded at the time of release of a variety are maintained during breeder seed production.

5. Plant Part Used

The plant part used for maintenance breeding differs from crop to crop. In wheat, barley and Triticale, spike or ear is used to grow further progeny, while in rice panicle is used for progeny testing. Details of plant parts used for maintenance breeding in different field crops are presented in Table 8.2.

TABLE 8.2: Plant Parts Used for Maintenance Breeding in different Field Crops

Sl.No.	*Name of Crop*	*Plant Part Used*	*Units Selected*	*Isolation m*
1	Wheat	Ear	300 to 500	5
2	Barley	Ear	300 to 500	5
3	Triticale	Ear	300 to 500	5
4	Maize	Ear	100 to 150	400
5	Pearl millet	Ear	500 to 1000	400
6	Sorghum	Panicle	200 to 500	200
7	Rice	Panicle	100 to 500	3
8	Rapeseed and Mustard			
	(i) Self-pollinated	Single Plant	400 to 500	50
	(ii) Cross-pollinated	Single Plant	2000 to 2500	100
9	Soybean	Single Plant	400 to 500	3
10	Groundnut	Single Plant	100 to 150	3
11	Sunflower	Single Plant	500 to 800	600
12	Sugarcane	Clump	100 Clumps	-
13	Cotton	Single Plant	200 to 250	50
14	Jute	Single Plant	500 to 600	50
15	Tobacco	Single Capsule	1 to 5	5
16	Potato	Single Plant	100 to 150	3
17	Black gram, Green gram	Single Plant	100 to 150	10
18	Chickpea, Field pea, Lentil	Single Plant	100 to 150	10
19	Pigeon pea	Single Plant	500 to 1000	200
20	Castor [CP]	Single Plant	200 to 500	1500 - 2000 m
21	Sesame	Single Plant	400 to 500	100

Sl.No.	Name of Crop	Plant Part Used	Units Selected	Isolation m
22	Linseed	Single Plants	300 to 400	50
23	Oats	Panicle	400 to 500	3
24	Niger	Single Plant	800 to 1000	400

6. Number of Plant Progenies Used

In wheat, the number of spike used for progeny testing may vary from 300 to 500 depending upon quantity of breeder seed to be produced. In Rice, 100 to 200 panicles are sufficient for progeny testing.

7. Isolation Distance

The isolation distance differs from crop to crop. It is lesser in self-pollinated species than in cross-pollinated species.

8. Selection of Field

Regions of moderate rainfall and humidity are more suited for seed production. Most of the field crops require a dry sunny period and moderate temperature for flowering and pollination. Excessive dew and rainfall lead to poor seed set, high pest and disease incidence, heavy seed loss and lodging at harvest. High temperature results in pollen desiccation at flowering, premature flowering and poor quality seeds.

Selection of field for seed plot should meet certain requirements as presented below:

1. The field should have good soil texture and fertility.
2. It should be free from volunteer plants and weed plants.
3. It should be free from soil borne diseases and insect pests.
4. The plot must not have the same crop in the previous season also.
5. The field should be well leveled with proper isolation.
6. The field should be free from destruction of potential weeds.

9. Crop Inspections

Crop inspections are carried out to ensure proper selection of field, proper isolation distance and crop health. Generally, three crop inspections, *viz.* before flowering, during flowering and at maturity are required.

10. Evaluation of Progenies

In the maintenance breeding, genetic purity, physical purity, germination and seed health are the main criteria taken in to account.

Maintenance Procedures

There are various breeding methods/approaches which are used for maintaining the genetic purity of varieties and parental lines of commercial hybrids.

In crop plants four methods, *viz.*, positive mass selection, negative mass selection, model bulk selection, and plant to row selection are used for maintaining varietal purity. These are discussed as follows:

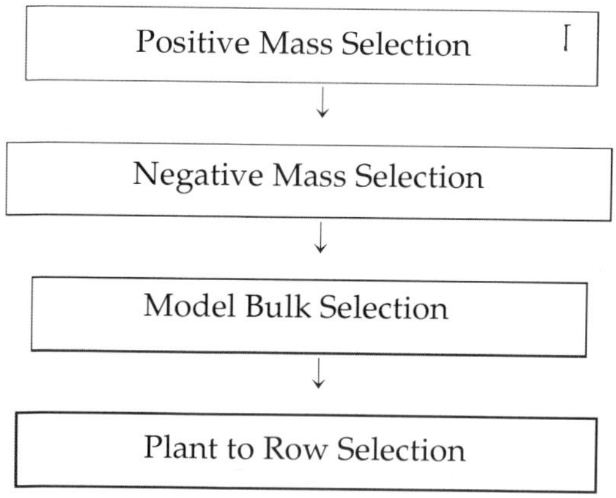

FIGURE 8.1: Methods of Maintenance Breeding.

1. Positive Mass Selection

It refers to selection of desirable plants [true to type] from the breeder seed field of a variety and mixing the seed of individual selected plants to grow the next generation. This is also known as bulk of large number of true to type plants. This method is widely used for varietal purification in various field crops. In cotton, this method was first used by Menbane in Texas, USA. Now this method is widely used for maintenance of varietal purity in cotton. Merits and demerits of this method are listed below.

Merits

It is a simple method of varietal maintenance. It is used for purification of improved varieties and parental lines of commercial hybrids. It retains diversity in the variety.

Demerits

Selection is based on phenotype which may not always be an indication of superior genotype. The progeny test is not carried out.

2. Negative Mass Selection

It refers to removal of undesirable plants [off types] from the field of a variety and mixing the seed of remaining plants to grow the next generation. In other words, it leads to removal of off type plants from the field of a variety. This method is widely used for maintaining varietal purity in various field crops. Merits and demerits of this method are presented as follows:

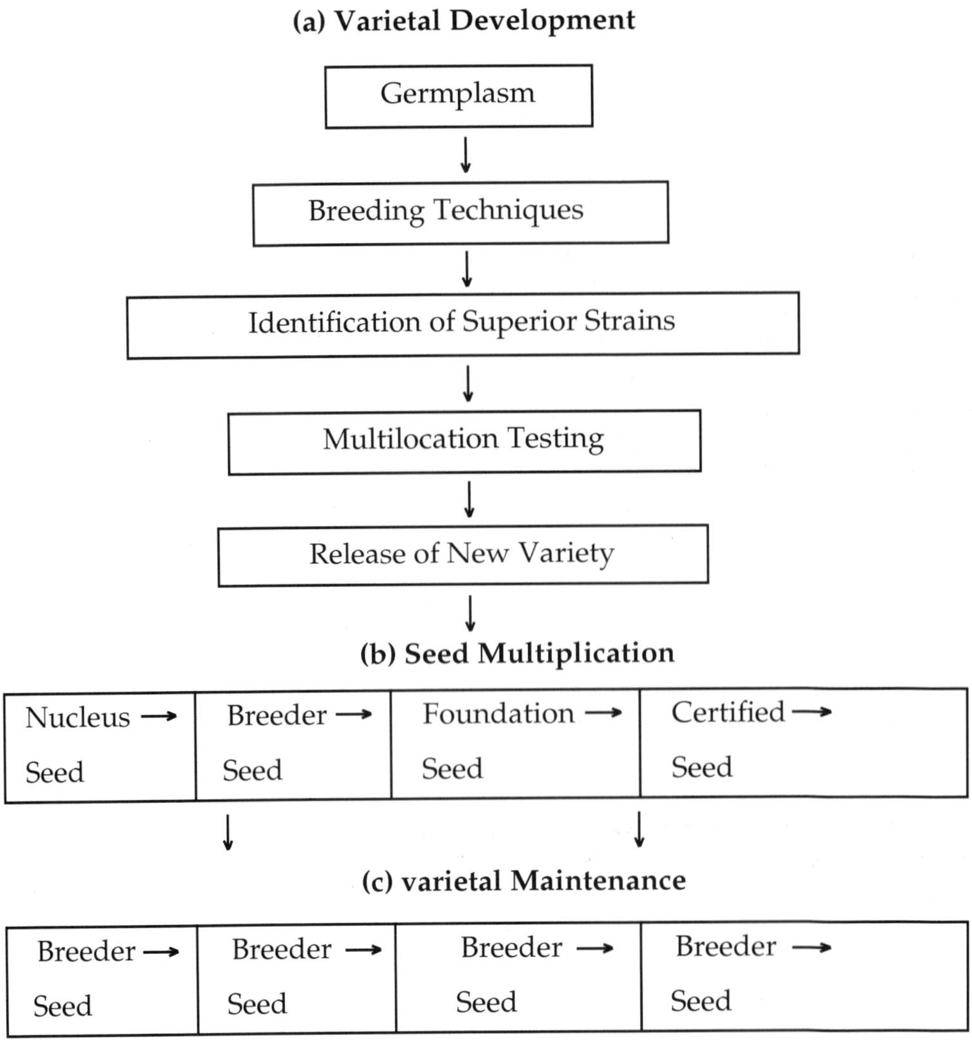

FIGURE 8.2: Plant Breeding Activities.

Merits

1. It is a simple method of varietal maintenance and is widely used for maintenance of varietal purity in different crops.
2. The maintenance procedure is easy and requires little skill.
3. The number of discarded plants is very few in proportion to plants retained.
4. The genetic diversity of the variety is not affected.
5. It is used both in self pollinated and cross pollinated species.

Demerits

1. The progeny test is not conducted
2. Sometimes, off type plants may escape which may lead to contamination of the variety in later cycles of seed multiplication.
3. It requires careful examination of off type plants.

TABLE 8.3: Comparison of Positive and Negative Mass selections

Sl.No.	*Particulars*	*Positive Mass Selection*	*Negative Mass Selection*
1	Plants selected/removed	Selection of true to the type plants	Removal of off type plants
2	Basis of selection	Phenotype	Phenotype
3	Progeny test	Not conducted	Not conducted
4	Applicable to	Both self and cross pollinated species.	Both self and cross pollinated species.
5	Genetic diversity	Retained	Retained
6	Base material used	Nucleus/breeder seed	Nucleus/breeder seed
7	Plants retained and bulked	Selected plants only	All plants after roguing

3. Model Bulk Selection

According to this method, large number of single plants [300-500] which are productive and true to the type are selected from the breeder seed plot of a variety. Then these plants are examined under laboratory conditions for the characteristics of the variety recorded at the time of its release. Those plants which conform to the characteristics of a variety are selected and their seed is bulked to grow nest generation. Rest of the plants are discarded. This method was first used by Manning (1955) in cotton. Now this method is widely used for maintenance breeding. Merits and demerits of this method are given below:

Merits

1. This is a simple method of varietal purification.
2. This is more scientific than earlier two methods.
3. It is based on both field and laboratory observations.
4. It is used in both self and cross pollinated species.

Demerits

1. Large population has to be grown for selection.
2. Selection is based on phenotype.
3. It requires more labor, space and time.

4. Plant to Row Method

This method of maintenance breeding is based on progeny test. It is also known as progeny row selection. In this method, large numbers of single plant selections

[500-1000] that are true to the type of a variety are selected from the breeder seed plot of the parent variety. In the next season, progeny rows of selected plants are grown. Those progeny that conform to the norms of variety are bulked and rest discarded. Several modifications of this method have been suggested by various researchers as follows:

(i) Some researchers are of the opinion that progeny test should be conducted at least for one year to assess the performance of the variety.

(ii) Others are of the opinion that one year progeny test is not sufficient. Progeny test should be conducted for two, three or more seasons. When several years of progeny test is carried out, it is known as pedigree system or pedigree method of selection. Merits and demerits of this method are presented as follows;

Merits

1. The selection is based on breeding value or progeny performance.
2. This is more scientific method than earlier three methods.
3. It involves both field and laboratory evaluation.
4. Progeny tests are based on large samples which provide more precision.

Demerits

1. It is more laborious and time consuming method.
2. It requires more experimental area for evaluation and testing.
3. It requires two or more crop season for testing.

Advantages of Maintenance Breeding

1. It prevents varieties from genetic deterioration and thus enhances life span of a variety.
2. It provides opportunities to study the efficacy of various maintenance procedures.
3. It helps in purification of improved varieties and parental lines of commercial hybrids.
4. It leads to quality seed production in the seed production chain.
5. The originality of the variety is maintained.

Limitations

There are some limitations of maintenance breeding which are briefly presented as follows:

1. Most of the testing procedures are based on phenotypic performance which is not always indication of superior genotypes.

2. In case of positive mass selection, major portion of plants is rejected that adds to the production cost.
3. Progeny row method requires long time [2-3 crop seasons] for evaluation of purity of a variety.
4. Progeny row method requires lots of experimental material and area for evaluation.
5. Since, large numbers of single plants have to be evaluated for key characters of a variety, only limited number of varieties can be handled at a time.

Summary

There are three important areas or activities of plant breeding, *viz.* (i) varietal development, (ii) seed multiplication and (iii) varietal maintenance. These have been explained. An area of plant breeding which deals with principles and methods of breeder seed production is referred to as maintenance breeding. It is also known as varietal maintenance technology.

There are various breeding methods/approaches which are used for maintaining the genetic purity of varieties and parental lines of commercial hybrids. In crop plants four methods, *viz.*, positive mass selection, negative mass selection, model bulk selection, and plant to row selection are used for maintaining varietal purity. Procedure, merits and demerits of these have been discussed. Advantages and limitations of maintenance breeding have been discussed.

Glossary

Maintenance breeding: An area of plant breeding which deals with principles and methods of breeder seed production; also known as varietal maintenance technology.

Model Bulk Selection: Selecting large number of single plants [300-500] from the breeder seed plot of a variety, examining these plants under laboratory conditions for the characteristics of the variety recorded at the time of its release and selecting those plants which conform to the characteristics of a variety and their seed is bulked to grow nest generation.

Negative Mass Selection: Removal of undesirable plants [off types] from the field of a variety and mixing the seed of remaining plants to grow the next generation.

Plant to Row Method: This method of maintenance breeding is based on progeny test. It is also known as progeny row selection. Progeny of single plant selections are planted, evaluated and progeny with true to the type are selected and their seed is mixed to grow the nest generation.

Positive Mass Selection: Selection of desirable plants [true to type] from the field of a variety and mixing the seed of individual selected plants to grow the next generation.

QUESTIONS

1. **Define maintenance breeding. Explain main points related to such breeding.**
2. **Describe briefly various method of maintenance breeding.**
3. **Explain briefly positive and negative mass selections. Give their merits and demerits.**
4. **What is model bulk selection? Describe its procedure, merits and demerits.**
5. **What is progeny row selection? Describe its procedure, merits and demerits.**
6. **Write short notes on the following:**

   (i) Positive Mass Selection (ii) Negative Mass Selection

   (iii) Model Bulk Selection (iv) Progeny Row Selection
7. **Explain briefly advantages and limitations of maintenance breeding.**

* * * *

Chapter 9

Genetic Basis of Maintenance Breeding

Introduction

Seed quality is determined by a number of genetic and physiological characteristics. The genetic component involves differences between two or more genetic lines, while differences between seed lots of a single genetic line comprise the physiological component. The genetic factors that can influence quality include genetic make-up, seed size and bulk density.

Deterioration in seed quality may begin at any point in the plant's development stage from fertilization onward. Seed quality depends upon the physical conditions that the mother plant is exposed during its growth stages, as well as harvesting, processing, storage and planting. Temperature, nutrients and other environmental factors also affect seed development and influence seed quality. High quality seeds are the result of proper production practices, which include:

(i) Proper maintenance of genetic purity

(ii) Good growing conditions

(iii) Proper timing and methods of harvesting

(iv) Precaution during processing, threshing, cleaning and drying.

(v) Appropriate seed storage and seed distribution systems.

Genetic Basis of Maintenance Breeding

The maintenance breeding or maintenance of varietal purity is based on certain genetic principles, which are referred to as genetic bases of maintenance breeding. Genetic bases of maintenance breeding include (1) control of seed source, (2) isolation distance, (3) rouging of off types, (4) seed certification and (5) grow out test. These aspects are briefly discussed as follows:

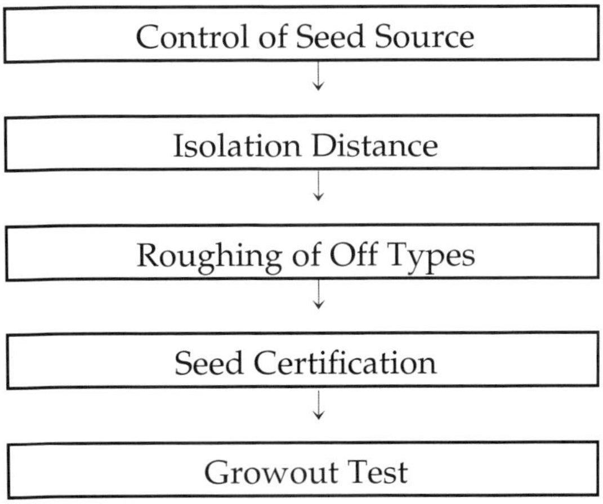

FIGURE 9.1: Genetic Bases of Maintenance Breeding.

1. Control of Seed Source

For quality seed production, the seed of improved category should be used and obtained from an approved source. Such seed helps in maintaining seed quality in the seed production chain. Four classes of seeds are generally recognized in seed certification chain namely breeder seed, foundation registered and certified. These classes are recognized by Association of official seed certifying agencies [AOSCA].

(i) **Breeder's seed:** It is a seed or vegetative propagating material which is directly controlled by originating or sponsoring breeder of institution and which provides increases of foundation seeds.

(ii) **Foundation seed**: It is the progeny of breeder seed which is used for the production of certified seed. Foundation seed is the source of all other certified seed classes, either directly or through registered seed.

(iii) **Registered seed:** It is the progeny of foundation which is used for the production of certified seed. This category of seed is omitted in India and directly certified seed is produced from the foundation seed.

(iv) **Certified Seed:** It is the progeny of foundation, registered or certified seed. This category of seed is distributed to the farmers for commercial crop production.

2. Isolation

The separation of the field of a variety from that of another variety of the same crop by a minimum prescribed distance to avoid contamination is called isolation. Isolation is essential to prevent pollination from unwanted pollen in the case of cross pollinated and often cross pollinated species to avoid mechanical mixture and chance cross pollination in self-pollinated crops. The isolation distance varies from 3m in self-pollinated crops like rice, wheat to 200m in maize, *etc.* It also varies with the type of seeds. In some cases like hybrid maize, the minimum isolation

distance may be considerably reduced by planting border rows of pollinator parent and by choosing a large field for seed production. Isolation is also required during harvesting, threshing and handling of seeds to avoid mechanical mixture. The isolation distance is different from crop to crop and from different classes of seeds. *i.e.* certified seeds and foundation seed plots.

TABLE 9.1: Minimum Isolation Distance for Foundation and Certified Seeds of Different Crops

Sl.No.	*Name of crop*	*Isolation Distance [Meters]*	
		Foundation Seed	*Certified Seed*
A	**Self-pollinated Species**		
1	Wheat	3	3
2	Barley	3	3
3	Rice	3	3
4	Oats	3	3
5	Groundnut	3	3
6	Blackgram	20	10
7	Greengram	20	10
8	Lentil	20	10
9	Khesari	20	10
10	Chickpea	20	10
11	Soybean	20	10
12	Field Pea	3	3
13	Cluster bean	20	10
14	French bean	50	25
15	Cowpea	50	25
16	Linseed	50	25
17	Tomato	50	25
18	Methi	50	25
B	**Cross-pollinated Species**		
1	Maize	400	200
2	Pearl millet	1000	200
3	Mustard/Rapeseed	400	200
4	Sunflower	400	200
5	Safflower	400	200
6	Cucurbits	800	400
7	Cabbage	1600	1000
8	Cauliflower	1600	1000
9	Radish/Turnip	1600	1000
10	Onoin	1600	400
11	Sugarbeet	1600	1000

Sl.No.	Name of crop	Isolation Distance [Meters]	
		Foundation Seed	Certified Seed
12	Castor	300	100
13	Carrot	1000	800
14	Palak	1600	1000
C	**Often Cross-Pollinated Species**		
1	Cotton	50	30
2	Sorghum	200	100
3	Pigeonpea	200	100
4	Brinjal	200	100
5	Okra	400	200
6	Chilli	400	200
7	Sesame	100	50

3. Rouging of Seed Fields

The process of removal of the off types plants from the field of an improved variety to avoid contamination is called roguing. The presence of off type [phenotypically different] plants is another source of genetic contamination. Their continued presence would certainly deteriorate the genetic purity of the variety. The removal of such type of plant is referred as rouging. There are three main sources of off- type plants as given below.

(i) The off-type plant may arise due to presence of recessive genes in heterozygous condition at the time of release of variety. (The recessive genes may also arise by mutation).

(ii) Off-type plants may also arise from volunteer plants or from seed produced by earlier crop.

(iii) Mechanical mixtures also leads to presence of off-type plants in a variety.

4. Seed Certification

Seed Certification is a legal system which ensures availability of high quality seed of improved variety to the farmers. The genetic purity in commercial seed production is maintained through a system of seed **certification.** The objective of seed certification is to make available crop seeds, tubers, bulbs, *etc.* of good quality to the farmers. Seed certification work is carried out by well experienced and qualified persons of seed certification agency. They carry out field inspection at appropriate stage of crop growth. They also make sure that the seed crop/seed lot is of the requisite genetic purity and quality. After crop harvest they ensure quality of the variety and at the processing plants they take samples for seed testing and also for grow-out-test. Certification requires two, three or four inspections depending upon the crop species.

5. Grow Out Tests

These tests are much useful in self-pollinated than in cross pollinated species. The authentic sample is planted after every 10 test samples for close comparison. Observations are made both on qualitative and quantitative traits of the test and authentic sample plots during the entire growing period. The frequency of various off types is recorded. This test is used to observe the genetic purity of the crop. Though it is more precise but is rarely used as it requires a much longer time and excellent green house facility. Varieties being grown for seed production should periodically be tested for genetic purity by grow – out – test, to make sure that, seed being maintained in their true form.

Hartmann and Kester (1968) have suggested following steps for maintaining genetic purity of the varieties in seed production chain.

1. Providing adequate isolation to prevent contamination by natural crossing or mechanical mixtures.
2. Rouging of seed fields, prior to the stage at which they could contaminate the seed crop.
3. Periodic testing of varieties for genetic purity.
4. Avoiding genetic shift by growing crops in areas of their adaptation only.
5. Certification of seed crops to maintain genetic purity and quality seed.
6. Adopting generation system. (the seeds produced is restricted to four generation only *i.e.* starting from breeders seeds.) and the seeds can be multiplied up to three more generations *i.e.* foundations, registered and certified.
7. Grow out Test: This test should be conducted to assess genetic purity during seed production.

Summary

Deterioration in seed quality may begin at any point in the plant's development stage from fertilization onward. Seed quality depends upon the physical conditions that the mother plant is exposed during its growth stages, as well as harvesting, processing, storage and planting. Temperature, nutrients and other environmental factors also affect seed development and influence seed quality.

The production of good quality seed is based on certain genetic principles as well as agronomic principles. Genetic principles include (i) control of seed source, (ii) isolation distance, (iii) rouging of off types, (iv) seed-certification and grow out test.

Glossary

Grow out Test: A test which is conducted to assess the genetic purity of a variety during seed production.

Isolation Distance: The separation of the field of a variety from that of another variety of the same crop by a minimum prescribed distance to avoid contamination.

Roguing: The process of removal of the off types plants from the field of an improved variety to avoid contamination.

Seed certification: A legal system which ensures availability of high quality seed of improved variety to the farmers.

Volunteer plants: The plants which grow in the field from previous crops.

QUESTIONS

1. **Describe briefly genetic principles involved in quality seed production.**
2. **Describe briefly agronomic principles involved in quality seed production.**
3. **Define isolation. Give ten examples of isolation distance for foundation and certified crops**
4. **Define roguing and describe its significance.**
5. **Define the following terms:**

   (i) Isolation distance (ii) Roguing

   (iii) Grow out Test (iv) Seed Certification
6. **Describe the significance of rouging and isolation distance in quality seed production.**
7. **Explain the role of seed certification in quality seed production.**

* * * *

Chapter 10

Agronomic Basis of Maintenance Breeding

Introduction

Agronomic principles which are related to maintenance breeding *i.e.* production of nucleus or breeder seed include (1) selection of field, (2) cultural practices, (3) plant protection, (4) weed control, (5) harvesting and threshing, (vi) cleaning, grading, testing and bagging. These aspects are briefly discussed as follows:

1. Selection of Field and Region

Regions of moderate rainfall and humidity are more suited for seed production. Most crops require a dry sunny period and moderate temperature for flowering and pollination. Excessive dew and rainfall lead to poor seed set, high pest and disease incidence, heavy seed loss and lodging at harvest. High temperature results in pollen desiccation at flowering, premature flowering and poor quality seeds.

Selection of field for seed plot should meet certain requirements. The field should have good soil texture and fertility. It should be free from volunteer plants and weed plants. It should be free from soil borne diseases and insect pests. The plot must not have the same crop in the previous season also. The field should be well leveled with proper isolation. The field should be free from destruction of potential weeds.

2. Cultural Practices

Standard agronomic practices such as sowing time, spacing, fertilizer-dose, hoeing and weeding are available for each crop. These cultural practices have to be strictly followed to raise a good seed crop. It is essential to harvest good yield of better quality of produce. Hence, recommended agronomic practices should be adopted for production of quality seeds. Proper crop rotation has to be adopted to

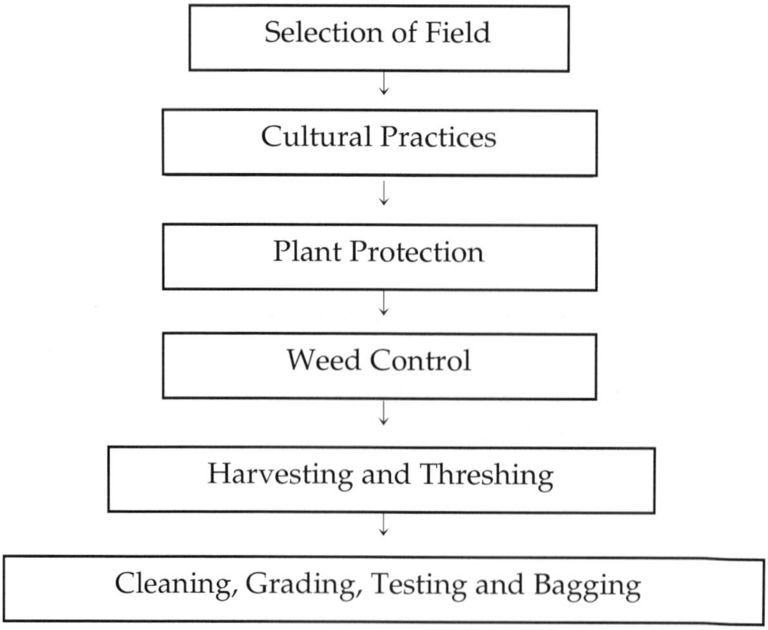

FIGURE 10.1: Agronomic Basis of Maintenance Breeding.

avoid contamination from volunteers of the same crop. The seed production should be taken up under irrigated conditions to ensure high yield and good quality.

3. Plant Protection

Insect pests and diseases may cause considerable damage to the crop reducing the yield and quality of seed. Incidence of some diseases may lead to the rejection of seed by the certifying agency as unfit for use. For example, maximum permissible limit for head smut in Sorghum is 1/1000. Hence, proper plant protection measures should be adopted to raise a healthy crop.

4. Weed Control

Effective weed control is a must for good seed production. Weeds reduce crop yield and their seeds contaminate the crop seed. It acts as host for number of pathogens. Some weeds are classified as noxious or objectionable weeds. Their seeds are similar to crop seeds and sometimes it is very difficult to separate them from seed of the concerned crops. It is difficult to eradicate such weeds once they infest a field. In addition they act as host of disease or pest.

5. Harvesting and Threshing

Considerable care should be taken to prevent mechanical mixture from other crops or weed seeds during harvesting and threshing. Threshing floor should be clean, while threshing. There should not be any damage to seeds.

6. Seed Processing

Seed processing includes drying, cleaning, grading, treatment with chemicals, bagging and labeling. The seed should be dried to appropriate moisture level to facilitate processing to prevent losses in germination and to reduce the chances of insect attack during storage. The seeds can be sun dried or can be dried by hot air.

Cleaning and Grading

Separation of inert matters weed seeds and seeds of other crops from seed is known as cleaning.

1. Specific gravity separators are used based on weight and size of seeds
2. Pneumatic separators : resistance to airflow
3. Spiral separators : Seed shape
4. Velvet roll: surface smoothness
5. Electronic separators: electrical properties
6. Electronic color separators: Seed color

Grading

Removal of smaller and shriveled seeds from the well filled healthy seeds is called grading. Air and screen machine is used for cleaning and grading.

Testing

After cleaning and grading the seed lots are tested for percentage of pure seed, weed seeds, seeds of other crops, inert matter and germination in the seed testing laboratory.

Treating

Before bagging, seeds are treated with suitable fungicides. It helps in controlling seed borne diseases and protects against damage by storage pest

Bagging

Seeds are distributed in bags of appropriate size.

Labeling

It contain details about the seed such as kind of seed, name of variety, purity, germination percentage, date of germination test, percentage of weed seeds, percentage of inert matter, name and address of the seller, period of validity of certification and treated with poison. The tag color is yellow, white and blue for breeder, foundation and certified seed, respectively.

Summary

The production of good quality seed is based on certain genetic principles as well as agronomic principles. Genetic principles include (i) control of seed source, (ii) isolation distance, (iii) rouging of off types, (iv) seed certification and grow out test.

Agronomic principles include (i) selection of field, (ii) cultural practices, (iii) plant protection, (iv) weed control, (v) harvesting and threshing, (vi) cleaning, grading and testing and bagging. These aspects have been amply discussed.

Glossary

Grow out Test: A test which is conducted to assess the genetic purity of a variety during seed production.

Isolation Distance: The separation of the field of a variety from that of another variety of the same crop by a minimum prescribed distance to avoid contamination.

Roguing: The process of removal of the off type plants from the field of an improved variety to avoid contamination.

Seed certification: A legal system which ensures availability of high quality seed of improved variety to the farmers.

Volunteer plants: The plants which grow in the field from previous crops.

QUESTIONS

1. **Describe briefly agronomic principles involved in quality seed production.**
2. **Define isolation. Give ten examples of isolation distance for foundation and certified crops**
3. **Define Roguing and describe its significance.**
4. **Define the following terms:**
   (a) Isolation distance
   (b) Grow out Test
   (c) Roguing
   (d) Seed Certification
5. **Describe the significance of rouging and isolation distance in quality seed production.**
6. **Explain the role of seed certification in quality seed production.**

* * * *

Chapter 11

Nucleus Seed Production of Cereals

Introduction

Important cereal crops include wheat, barley, triticale and rice. A brief account of distribution, cultivated species, origin and evolution, reproduction and pollination, breeding objectives, breeding procedures, breeding centres, practical achievements and nucleus seed production related to wheat, barley, triticale and rice is presented in this chapter.

1. Wheat (*Triticum aestivum* L.)

General Information

It includes, family, global distribution, national distribution, types, uses as briefly presented below:

(i) **Family**: Wheat belongs to the family Poaceae (old *Gramineae*). It is an important cereal crop of cool climates.

(ii) **Global Distribution**: It is widely grown the world over and stands first among cereals in area and production. Wheat is cultivated in USA, U.K. Russia, Ukraine, China, Japan, Argentina, Canada, Mexico, India, Pakistan and many other countries.

(iii) **National Distribution:** In India, Uttar Pradesh, Madhya Pradesh, Punjab, Haryana, Rajasthan, Bihar and Gujarat are the major wheat growing states.

(iv) **Types:** Wheat is classified on various bases. Based on consumption it is classified as bread wheat, common wheat or soft wheat (*T. aestivum*) and hard wheat or macaroni wheat (*T. durum*). Based on cold tolerance it is of two types, *viz.* winter wheat (which can withstand chilling temperature) and spring wheat.

(v) **Uses:** Wheat is used for human consumption in a variety of products such as bread, cakes, noodles, cookies, chapati, macaroni, *etc.*, in various countries.

Cultivated Species

There are two cultivated species of wheat, *viz.* common wheat (*Triticum aestivum*) and durum wheat (*T. turgidum* L.). Common wheat is hexaploid ($2n = 6x = 42$), whereas durum wheat is tetraploid ($2n = 4x = 28$). The former is more widely adapted than the latter. Common wheat is used for bread, cakes, noodles, cookies, chapati, *etc.*, whereas the durum wheat is used mainly for macaroni and some flat bread. There are 16 wild species of wheat, out of which six are diploids ($2n = 14$), seven tetraploids ($2n = 4x = 28$), and three hexaploids ($2n = 6x = 42$). Wild species are used in hybridization programs for transfer of resistance to biotic and abiotic factors, adaptation and other desirable characters into cultivated species.

Origin and Evolution

Near East is the center of origin of bread wheat. Genetic investigations have established that bread wheat has evolved from crosses among three species of wheat in nature in two steps [belonging to A, B and D genome] as follows:

(a) Einkorn wheat [*T. monococcum* x unknown species] = Emmer wheat [T.turgidum]

(b) Emmer wheat x Einkorn wheat [*T. tauschi*] = bread wheat [*T. aestivum*]

Durum wheat has originated from a cross between two diploid wild species.

(i) Origin of tetraploid wheat [*Triticum turgidum*]

Parents	*Triticum monococcum*	x Unknown species
Genome	AA [2n=14]	BB [2n=14]
F_1	AB [2n=14] Sterile	
Chromosome doubling in nature	AABB [2n =28] *Triticum turgidum* [Tetraploid-Fertile]	

(ii) Origin of hexaploid wheat [*Triticum aestivum*]

Parents	*Triticum turgidum*	*x Triticum tauschi*
Genome	AA BB [2n=28]	DD [2n=14]
F_1	ABD [2n=21] Sterile	
Chromosome doubling in nature	AABBDD [2n =42] *Triticum aestivum* [Hexaploid-Fertile]	

FIGURE 11.1: Probable Origin of Hexaploid Bread Wheat [*Triticum aestivum*].

The chromosome doubling took place in nature in above crosses. Other two species of hexaploid wheat, *viz. T. compactum, T. spherococcum* originated through spontaneous mutation of *T. aestivum*. The durum wheat probably originated from cultivated Emmer wheat [*T. turgidum* var. *dicoccum*] after several spontaneous mutations.

Reproduction and Pollination

Wheat is a self-pollinated and seed propagated crop. Main points are presented as follows:

(i) Flowers: Flowers are bisexual. Flowers open generally after pollination is over (chasmogamy). This chamogamy protomes autogamy. The flower is covered by two membraneous structures. The outer is called lema and inner as palea.

(ii) Pollination: It is self-pollinated crop. Cross pollination is less than 1 per cent.

(iii) Propagation: Sexually produced seeds are used for propagation.

(iv) Tillering: Tillering is a common feature of wheat crop.

(v) Isolation Distance: An isolation distance of 3 metres is absolutely safe for the production of breeder and foundation seed.

Breeding Objectives

In wheat, major breeding objectives are higher grain yield, better quality, early maturity and wider adaptability and resistance to biotic (diseases and insects) and abiotic (drought, salinity, lodging, *etc.*) stresses.

(i) Higher grain yield: Major yield components are: ear length, kernel per ear, number of productive tillers per plant and kernel size.

(ii) Quality: Grain quality includes color, size and lusture, protein content, lysine content, *etc.*

(iii) Disease Resistance: Major diseases include rusts (black or stem rust, brown or leaf rust and yellow or stripe rust), loose smut, hill bunt, Karnal bunt, powdery mildew and leaf blight.

(iv) Insect Resistance: In India insects are not the serious problem. In USA, Hessian fly, stem sawfly and green bugs are major insect pests of wheat.

Breeding Procedures

Breeding procedures which are used for wheat improvement are of three types, *viz.* general methods, special methods and population improvement procedures as follows:

(i) General Methods: These methods include introduction, pure line selection, pedigree, back cross and multiline breeding, bulk and single seed descent methods.

(ii) Special methods: Such methods include: mutation breeding, distant hybridization, heterosis breeding and transgenic breeding.

(iii) Population Improvement Procedures: Such procedures include recurrent selection and diallel selective mating systems. These are used for specific purposes.

Breeding Centers

There are two types of wheat breeding centers as follows:

(i) International Centers: There is only one International wheat breeding center, *i.e.* International Maize and Wheat Improvement Centre (CIMMYT), Mexico. International multi-locational testing is carried out by CIMMYT for identification and release of varieties for different countries. The global gene poll of wheat is also maintained at this centre.

(ii) National Centers: In India, wheat breeding work is carried out by following organizations:

(a) Indian Institute of Wheat and Barley Research, Karnal. The new varieties are released through coordinated project after multi-location testing for 3-5 years.

(b) State Agricultural Universities located in wheat growing regions.

(c) Some private seed companies.

Practical Achievements

Remarkable work on wheat breeding has been done by CIMMYT which led to green revolution in wheat production the world over. The semi-dwarf varieties of wheat have been developed through the use of Japanese line Norin 10 as a source of dwarfing gene. The semi dwarf varieties developed by CIMMYT at Mexico have spread to different wheat growing countries resulting in revolution in wheat production. The productivity of semi-dwarf varieties is about two and half times more than old tall growing varieties. Moreover, semi-dwarf varieties are highly resistant to lodging and are highly responsive to fertilizer doses. In India, several high yielding varieties of wheat have been released for different states through coordinated project. Some high yielding varieties of common wheat for different situations are listed below:

1. **Irrigated timely sown**: HD 2329, HD 2009, HD 2428, PBW 154, CPAN 3004 (Punjab, Haryana, Western U.P., Rajasthan)
2. **Irrigated late sown**: HD 2285, HD 2270, PBW 226, PBW 138 (Above 4 states)
3. **Rain-fed timely sown**: C 306, WL 410, IWP 72, PBW 65, PBW 175, WL 2265 (Above States)
4. **Salt affected soils:** WH 157, KRL 1-4
5. **Hilly Regions:** CPAN 1796, HD 208, HD 2380, HS 240, UP 1109 (U.P., J&K and H.P.) Varieties of durum wheat are PBW 34, PBW 215 and Raj 1555 for irrigated timely sown areas and JK 12 for rain-fed timely sown areas.

Nucleus Seed Production

Nucleus seed is the base material in seed production chain. The important steps involved in the nucleus seed production of wheat, barley and Triticale are briefly presented as follows:

(i) **Base Material:** Nucleus seed is produced from nucleus seed or breeder seed or sometimes from foundation seed. Select good quality nucleus or breeder or foundation seed for production of nucleus seed. Grow the breeder seed in large plot of at least 0.25 acre. The field of breeder seed production already planted can be used for production of nucleus seed. There is no need of planting separate field for such purpose.

(ii) **Breeding Method Used**: In wheat Ear to row selection is used for production of nucleus seed. One Ear or Spike true to the type is selected per plant at maturity. Seed of each selected Ear is harvested separately and stored in separate envelope.

(iii) **Number of Ear to be selected**: It depends on the quantity of nucleus seed to be produced. Generally, 300 to 500 Ears true to type of the variety are selected and seed of each selected Ear is harvested separately.

(iv) **Planting Progeny Rows**: In the second crop season progeny row of each selected Ear is planted separately. Each Ear to row progeny is carefully examined from ear emergence onwards. Off type plants are removed from each progeny row. If variants in a progeny row are many, the entire progeny row is rejected.

(v) **Laboratory Evaluation:** In the laboratory, seed of all progeny of single plants are evaluated in terms of shape, size, colour and lustre.

(vi) **Seed Bulking:** The seed of all true to the type progeny rows is bulked. This constitutes nucleus seed, which is used for production of breeder seed or nucleus seed in the third crop season. The bulk seed is grown in larger plots to produce breeder seed for further use in seed production chain.

TABLE 11.1: Steps in Nucleus Seed Production in Wheat, Barley and Triticale

Sl.No.	*Season*	*Particulars*	*Details*	*Remarks*
1	First	Base Population to be raised	0.25 Acre	In disease free field
		Single plants to be selected	300-500	Tag selected plants
2	Second	Ear rows of single plants	300-500	In isolation of 3m
		Laboratory evaluation of Ear row progeny	200	For seed characters
		Mixing true to type progeny rows	150	This constitutes nucleus seed
3	Third	Breeder seed production		Using nucleus seed

Note: Keep some nucleus seed in reserve for use in case of crop failure.

2. Rice (*Oryza sativa* L.)

General Information

(i) Family: Rice belongs to the family *Poaceae* [earlier *Gramineae*]. It is an important cereal food crop of global significance. It It is grown in humid tropical and subtropical climates.

(ii) Global Distribution: The major rice growing countries are China, India, Japan, Pakistan, Southeast Asia and Brazil. Rice is one of the principal food crops of the world.

(iii) National Distribution: In India, rice is grown in almost all the states.

(iv) Types: Rice is classified on the basis of fineness of grain as fine, and coarse; on the basis of scent as scented and non-scented.

(v) Uses: Rice is consumed as human food in a variety of ways, such as pullao, idle, plain rice after boiling, for making dosa and several other food preparations.

Cultivated Species

There are two cultivated species of rice, *viz.* Asian rice (*Oryza sativa* L.) and African rice (*Oryza glaberrima*). Both the species are diploid ($2n = 24$). Asian rice is the predominant species which has spread to different parts of the world. The African rice is still confined to tropical west Africa. There are three races of Asian rice, *viz.* Indica, Japonica and Javanica. Indica is grown in tropical climate, Japonica in the temperate region and the race Javanica is intermediate between Indica and Japonica. There are 22 wild species of rice which are found in the tropics of the world. Out of these, 8 are tetraploid species ($2n = 4x = 48$).

Origin and Evolution

Rice has originated in Asia as well as in Africa. The Asian rice [*Oyiza sativa*] has originated from perennial Asian wild species [Oryza rupipogan] and the African cultivated rice [*Oyiza glaberrema*] is believed to have originated in West Africa from wild diploid species [*Oryza breviligulata*]

Reproduction and Pollination

Rice is a self-pollinated and seed propagated crop. Main points are presented as follows:

(i) Flowers: Flowers are bisexual.

(ii) Pollination: Rice is self-pollinated crop. Cross pollination is less than 1 per cent.

(iii) Propagation: Sexually produced seeds are used for propagation.

(iv) Tillering: Tillering is a common feature of rice crop.

(v) Isolation Distance: An isolation distance of 3 meters is absolutely safe for the production of breeder and foundation seed.

Breeding Objectives

In rice, major breeding objectives are higher grain yield, better quality, early maturity and wider adaptability and resistance to biotic stresses as follows:

(i) **Higher grain yield:** Major yield components are: panicle length, number of grains per panicle, number of productive tillers per plant and test weight.

(ii) **Grain quality:** Grain quality includes color, size and texture, protein content, lysine content, fragrance, *etc.* Cooking qualities include kernel elongation, volume expansion, water absorption, gel consistency, *etc.*

(iii) **Disease Resistance:** Major diseases include blast, bacterial blight, stem rot, brown spot, virus disease, *etc.*

(iv) **Insect Resistance:** Major insects include, stem borer, brown hopper, Gandhi bug and Gall fly.

Breeding Procedures

Breeding procedures which are used for rice improvement are of three types, *viz.* general methods, special methods and population improvement procedures as follows:

(i) **General Methods**: These methods include introduction, pure line selection, pedigree, back cross and multiline breeding, bulk and single seed descent methods.

(ii) **Special methods**: Such methods include: mutation breeding, tissue culture, heterosis breeding and transgenic breeding.

(iii) **Population Improvement Procedures**: Such procedures include recurrent selection and diallel selective mating systems. These are used for specific purposes.

Breeding Centers

There are two types of rice breeding centers as follows:

(i) **International Centers:** There is only one International rice breeding center, *i.e.* International Rice Research Institute, Philippines. International multi-locational testing is carried out by IRRI for identification and release of varieties for different countries. The global gene poll of rice is also maintained at this centre.

(ii) **National Centers:** In India, rice breeding work is carried out by following organizations:

(a) National Rice Research Institute, Cuttack, Odisha.

(b) Indian Institute of Rice Research, Hyderabad, Tenangana.

(c) State Agricultural Universities located in rice growing regions.

(d) Some private seed companies.

Practical Achievements

IRRI has played significant role in the genetic improvement of rice. Several semidwarf, high yielding, varieties with wider adaptation have been released by IRRI for different countries. The Chinese Indica line Dee Geo-woo-gen has been used as a source of dwarfing gene in evolving semi-dwarf varieties of rice. Evolution of semi-dwarf varieties resulted in green revolution in rice. Important rice varieties developed by IRRI include, IR 8, IR 20, IR 22, IR 24, IR 26, IR 28, IR 29, IR 30, IR 32, IR 34 and IR 36. Out of these, IR 8, IR 20 and IR 36 have been released for India. In China, hybrid rice using cytoplasmic genetic male sterility system has been developed for commercial cultivation. Hybrid rice has also been developed in India by Indian Institute of Rice Research, (IIRS) Hyderabad. However, its commercialization will take some more time. In India, more than 100 improved varieties of rice for different states have been released by NRRI, [Cuttack], IIRS [Hyderabad] and State Agricultural Universities. The bacterial blight resistant varieties developed by IRRI and salinity resistant, scented and high yielding varieties developed for cultivation in India are listed below:

1. **Bacterial blight resistant varieties:** IR 20, IR 22, IR 26, IR 28, IR 29, IR 30, IR 32 and IR 34.
2. **Salinity resistant varieties**: Usar Dhan 1 (U.P.), PVR 1 (Tamil Nadu), MCM 1, MCM 2 (A.P.) Bhurarata, Kalarata (Maharashtra), Korekagga, Bilikagga (Karnataka), Damodar, Getu, Dosal, Nanobokhra, Nonasail, Rupsail (West Bengal), Pokkai and Chotupokkali (Kerala).
3. **Scented varieties**: Type 3, N 12, N 10 (B), Kasturi, Pusa, Basmati.
4. **High yielding varieties:** IR 8, IR 20, IR 36, Vikas, Savitri, Anamika, Mansarovar, *etc.*

Nucleus Seed Production

Nucleus seed is the base material in seed production chain. The important steps involved in the nucleus seed production of rice are briefly presented as follows:

(i) **Base Material**: Nucleus seed is produced from nucleus seed or breeder seed or sometimes from foundation seed. Select good quality nucleus or breeder or foundation seed for production of nucleus seed. Grow the breeder seed in large plot of at least 0.25 acre. The field of breeder seed production already planted can be used for production of nucleus seed. There is no need of planting separate field for such purpose.

(ii) **Breeding Method Used**: In rice, panicle to row selection is used for production of nucleus seed. From each plant one panicle true to the type is selected at maturity. Seed of each selected panicle is harvested separately and stored in separate envelope.

(iii) **Number of Panicle to be selected**: It depends on the quantity of nucleus seed to be produced. Generally, 100 to 500 panicles true to type of the variety are selected and seed of each selected panicle is harvested separately.

(iv) Planting Progeny Rows: In the second crop season progeny row of each selected panicle is planted separately. An isolation distance of 3m is required all around the progeny rows. Each panicle to row progeny is carefully examined at different stages of crop growth and off type plants are removed from each progeny row. If variants in a progeny row are many, the entire progeny row is rejected.

TABLE 11.2: Steps in Nucleus Seed Production of Rice

Sl.No.	*Season*	*Particulars*	*Details*	*Remarks*
1	First	Base Population to be raised	0.25 Acre	In disease free field
		Panicles to be selected	200	Tag selected plants
2	Second	Panicle rows of single plants	200	In isolation of 3m
		Laboratory evaluation of panicle row progeny	200	For seed characters
		Mixing true to type progeny rows	150	This constitutes nucleus seed
3	Third	Breeder seed production		Using nucleus seed

Note: Keep some nucleus seed in reserve for use in case of crop failure.

(v) Laboratory Evaluation: In the laboratory, seed of all progeny of single plants are evaluated in terms of shape, size, colour and lustre.

(vi) Seed Bulking: The seed of all true to the type progeny rows is bulked. This constitutes nucleus seed, which is used for production of breeder seed or nucleus seed in the third crop season. The bulk seed is grown in larger plots to produce breeder seed for further use in seed production chain.

(vi) Grow out Test: This test is conducted to examine genetic purity.

Summary

Important cereal crops include wheat, barley, triticale and rice. An area of plant breeding which deals with principles and methods of breeder seed production is referred to as maintenance breeding or varietal maintenance technology. Main points related to purpose, base material, breeding procedure used, main focus, plant part used, progenies grown, isolation distance, selection of field, crop inspection and evaluation of selected progenies for maintenance breeding [nucleus seed production] of cereal crops have been discussed.

QUESTIONS

1. **Describe briefly genetic principles involved in quality seed production.**
2. **Describe briefly procedure of breeder seed production in wheat.**
3. **Define Roguing and describe its significance.**

4. **Define the following terms:**
   (i) Isolation distance
   (ii) Roguing
   (iii) Grow out Test
   (iv) Seed Certification
5. **Describe the significance of rouging and isolation distance in quality seed production.**
6. **Explain the role of seed certification in quality seed production.**
7. **Describe briefly the procedure of nucleus seed production in Wheat.**
8. **Explain briefly the procedure of nucleus seed production in Barley.**
9. **Discuss briefly the procedure of nucleus seed production in pearl Rice.**

* * * *

Chapter 12

Nucleus Seed Production of Millets

Introduction

Important millet crops include maize, *Sorghum* and pearl millet. An area of plant breeding which deals with principles and methods of breeder seed production is referred to as maintenance breeding or varietal maintenance technology. A brief account of distribution, cultivated species, origin and evolution, reproduction and pollination, breeding objectives, breeding procedures, breeding centers, practical achievements and nucleus seed production related to maize, *Sorghum* and pearl millet is presented in this chapter.

1. Maize (*Zea mays* L.)

General Information

(i) **Family:** Maize belongs to the monocot family *Gramineae* (*Poaceae*).

(ii) **Global Distribution**: Maize is grown in Central and South America, Africa, Central Europe and Asia.

(iii) **National Distribution:** In India, U.P., Bihar, Rajasthan, M.P., Punjab and H.P. are the major maize growing states.

(iv) **Types:** Maize is of two types, *viz.* flints (with round seeds) and dents (with flat seeds). The flour can be easily made from flints than from dents. Maize has C4 photosynthetic pathway and is more efficient than C3 plants.

(v) **Uses**: Maize is an important cereal crop of global importance, which is used for human consumption in developing and undeveloped countries and as livestock feed in developed countries. Moreover, it is also used as green fodder for livestock.

Cultivated Species

There are five species of the genus Zea *viz. Zea maxicana, Z. perennis, Z. luxurian,*

Z. diploperenis and *Z. mays*. The first four species are wild and commonly called *Teosinte*. The last species *i. e, Z. mays* is the only cultivated species. The teosinte species are easily crossed with maize. The diploid chromosome number in maize is 20. The wild species, *Z. perennis* is tetraploid ($4n = 40$); rest are diploids. The other genus which is close to maize is *Tripsacum*. However, special techniques are required to cross maize and *Tripsacum*. There are 16 species of *Tripsacum* in which the somatic chromosome number varies from 36 to 108. The tetraploid wild species *Z. perennis* is believed to have originated from *Z. mexicana*.

Origin and Evolution

Central America (Mexico) is the place of origin of maize. The exact genetic origin of maize is not known so far. There are four different views (hypotheses) about the possible genetic origin of maize which are summarized below:

1. Maize, *Teosinte* and *Tripsacum* have originated from a common ancestor (from high lands of Mexico and Guatemala) which is now extinct (Weather wax, 1955).
2. Maize originated probably from a cross between Coix and *Sorghum* each with 10 chromosomes (Anderson, 1945).
3. Modern maize is believed to have originated from crosses between wild maize (pod corn) and *Tripsacum*. The *Teosinte* probably developed from a cross between cultivated maize and *Tripsacum* in Central America (Mangelsdorf and Reeves, 1939).
4. Some workers believe that maize has derived from *Teosinte* by direct selection (Beadle, 1939).

Reproduction and Pollination

Maize is a cross-pollinated and seed propagated crop. Main points are presented as follows:

(i) **Flowers:** Maize is a monoecious plant. Male and female flowers are separate on the same plant. Male inflorescence is called tassel and female flower is known as silk. Tassel is present on the top and silk in the middle portion of the plant.

(ii) **Cross Pollination:** Under natural condition, it is more than 99 per cent. Pollen is very light and cross pollination occurs by wind.

(iii) **Dichogamy**: In maize male and female flowers mature at different time which promotes cross pollination.

(iv) **Propagation**: Maize is a seed propagated crop.

(v) **Tillering**: Maize is a non-tillering cereal.

(vi) **Isolation Distance**: An isolation distance of 400 m and 200 m is required for the production of foundation and certified seed of open pollinated varieties.

Breeding Objectives

In maize, major breeding objectives are higher grain yield, better quality, early maturity and wider adaptability and resistance to biotic (diseases and insects) and abiotic (drought, salinity, lodging, *etc.*) stresses as follows:

(i) **Higher grain yield:** Major Yield Components are: ear length, kernel per row and per ear, size of kernel and number of ear per plant.

(ii) **Grain Quality**: Grain quality includes grain color, protein content, lysine and tryptophan. The gene opaque 2 is the source of high lysine content. The protein found in endosperm is known as Zein and is deficient in lysine and tryptophan.

(iii) **Disease Resistance**: Major diseases are root rot, stalk and ear rot, leaf blight, leaf spot, downy mildew and mosaic virus.

(iv) **Insect Resistance**: In India stem borer is the major insect pest.

Breeding Procedures

Breeding approaches which are used for maize improvement are of three types, *viz.* population improvement, inbred development and cultivar development.

(i) **Population Improvement Methods**: These methods include mass selection, progeny selection or ear to row selection, recurrent selection, disruptive mating and selection and bi-parental mating.

(ii) **Inbred Improvement Methods**: Such procedures include back crossing, convergent improvement and gamete selection. These are used for improving inbreds for various economic traits,

(iii) **Methods for Cultivar Development**: Mass selection is used for development of open pollinated cultivars and their maintenance. Heterosis breeding is used for developing high yielding hybrids. In maize three types of hybrids are developed, *viz.* single cross hybrids, three way cross hybrids and double cross hybrids. Synthetic and composite varieties are also developed to partially exploit heterosis. Tissue culture helps in developing haploids and creating additional variability.

Breeding Centers

There are two types of maize breeding centers as follows:

(i) **International Centers:** There is only one International maize breeding center, *i.e.* International maize and Wheat Improvement Centre (CIMMYT), Mexico. International multi-locational testing is carried out by CIMMYT for identification and release of varieties for different countries. The global gene poll of maize is also maintained at this center.

(ii) **National Centers:** In India, maize breeding work is carried out by following organizations:

(a) Indian Institute of Maize Research, Ludhiana. The new varieties are released through coordinated project after multi-location testing for 3-5 years.

(b) Vivekanand Parvatiya Krishi Anusandhan Shala, Almora, especially for hill regions.

(c) State Agricultural Universities located in maize growing regions.

(d) Some private seed companies.

Practical Achievements

In maize, several composite and hybrid cultivars have been released by AICMIP for various agro-climatic conditions of the country. Some important high yielding composite and hybrid cultivars with wider adaptability are listed below:

1. **Composite:** Jawahar, Vijay, Kisan, Amber, Sona, Vikram, Protina, Rattan, Shakti, Vikas, Navjot, Ageti 76, Kanchan, Diara 3, D 765, MCU 508, Kiran, Surya, Tarun, Arun, Renuka, *etc.*
2. **Hybrids:** Ganga safed, Ganga 5, Ganga 9, Deccan 103, Sartaj Makka 1, VL 42, *etc.* Hybrids Him 123 and Him 128 have been released for the hilly regions of UP, HP, Assam and Sikkim. Both genetic and cytoplasmic genetic male sterility systems are available in maize. Usually, CGMS system is used for the development of single, three-way and double cross hybrids in maize.

Nucleus Seed Production

Nucleus seed is the base material in seed production chain. The important steps involved in the nucleus seed production of maize are briefly presented as follows:

(i) Base Material: Nucleus seed is produced from nucleus seed or breeder seed or sometimes from foundation seed. Select good quality nucleus or breeder or foundation seed for production of nucleus seed. Grow the breeder seed in large plot of at least 0.25 acre. The field of breeder seed production already planted can be used for production of nucleus seed. There is no need of planting separate field for such purpose.

(ii) Breeding Method Used: In maize, Ear to row selection is used for production of nucleus seed. One Ear true to the type is selected per plant during silk emergence and selfed. Seed of each selected Ear is harvested separately and stored in separate envelope.

(iii) Number of Ear to be selected: It depends on the quantity of nucleus seed to be produced. Generally, 100 to 150 Ears true to type of the variety are selected, selfed and seed of each selected Ear is harvested separately.

(iv) Planting Progeny Rows: In the second crop season progeny row of each selected Ear is planted separately adopting 400m isolation distance. Each Ear to row progeny is carefully examined from ear emergence onwards. Off type plants are removed from each progeny row. If variants in a progeny row are many, the entire progeny row is rejected.

(v) **Laboratory Evaluation:** In the laboratory, seed of all progeny of single plants are evaluated in terms of shape, size, colour and lustre.

(vi) **Seed Bulking:** The seed of all true to the type progeny rows is bulked. This constitutes nucleus seed, which is used for production of breeder seed or nucleus seed in the third crop season. The bulk seed is grown in larger plots to produce breeder seed for further use in seed production chain.

TABLE 12.1: Steps in Nucleus Seed Production of Maize

Sl.No.	*Season*	*Particulars*	*Details*	*Remarks*
1	First	Base Population to be raised	0.25 Acre	In disease free field
		Ears to be selected	500	Tag selected ears
2	Second	Plant Ear to rows progeny	500	In isolation of 400m
		Laboratory evaluation of Ear row progeny	250	For seed characters
		Bulking equal number of seed from best Ear/cob	200	This constitutes nucleus seed
3	Third	Breeder seed production		Using nucleus seed

Note: Keep some nucleus seed in reserve for use in case of crop failure.

2. Sorghum (*Sorghum bicolour* Moench)

General Information

(i) **Family:** *Sorghum* belongs to the family *Gramineae.* It is an important food crop of global significance which is cultivated in tropical and subtropical climates especially in the semiarid tropics the world over. *Sorghum* is a C4 plant and, therefore, has high photosynthetic efficiency.

(ii) **Global Distribution**: The major *Sorghum* producing countries are Africa, India, China, Pakistan, USA, Australia, Argentina and Mexico.

(iii) **National Distribution**: In India, major *Sorghum* growing states are Maharashtra, Gujarat, Tamil Nadu, Karnataka, Rajasthan and Madhya Pradesh.

(iv) **Types**: It is of two typed, *viz.* grain sorghum and fodder sorghum.

(v) **Uses:** The grain is used as human food in various ways and stalk as animal feed. In south India, where winter is mild, it is also grown in rabi season.

Cultivated Species

There is only one cultivated species of *Sorghum, i.e. Sorghum bicolour* which is diploid ($2n = 2x = 20$). In *Sorghum,* the panicle has wide variability ranging from very open and loose to compact. Based on panicle morphology, there are five races of *Sorghum, viz.* Bicolour (open panicle and elongated grains), Guinea (open panicle and flattened grains), Caudatum (high yield and bright seed colour), Kafir (Semi compact to compact panicle and spherical grain), and Durra (round grains). The races

Kafir and Durra have several useful genes. Stem juice of fodder *Sorghum* is sweeter than grain *Sorghum*. The important wild species are *S. bicolour* spp. *arundinaceum* ($2n = 20$) and *S. halepense* ($2n = 4x = 40$).

Origin and Evolution

It is generally accepted that Ethiopia is the centre of origin of cultivated *Sorghum*. From there it spread to other parts of the world. The diploid wild species *S. arundinaceum* ($2n = 20$) is considered to be the likely ancestor of cultivated *Sorghum*. This wild species is still found growing in Ethiopia and its adjacent regions.

Reproduction and Pollination

The cultivated *Sorghum* is seed propagated and self-pollinated. Main points are presented as follows:

(i) **Pollination**: It is a self-pollinated crop. Natural cross pollination varies from 5 to 30 per cent depending upon genotype and environmental conditions. Pollen is very light and cross pollination occurs by wind.

(ii) **Protogyny**: It is common which promotes cross pollination.

(iii) **Propagation**: Sorghum is a seed propagated crop.

(iv) **Tillering**: In Sorghum, tillering is important.

(v) **Isolation Distance**: An isolation distance of 200 m. and 100 m. is essential for the production of foundation and certified seed.

Breeding Objectives

In *Sorghum*, major breeding objectives are higher grain yield, better quality, early maturity and wider adaptability and resistance to biotic (diseases and insects) and abiotic (drought and lodging) stresses as follows:

(i) **Higher grain yield**: Major Yield Components are: panicle size and type, grains per panicle, productive tillers per plant and test weight.

(ii) **Quality**: Quality characters include chapatti and bread making qualities, protein and lysine content.

(iii) **Disease Resistance**: Major diseases are leaf blight, zonate leaf spot, downy mildew, rust, ergot or sugary disease and charcoal rot.

(iv) **Insect Resistance**: In India stem borer, stem fly and midge are important insects. *Striga* is a root parasite weed in *Sorghum*.

Breeding Procedures

Breeding approaches which are used for Sorghum improvement are of two types as presented below:

(i) **General Methods**: These methods include introduction, pure line selection, pedigree breeding, back crossing and composite breeding.

(ii) **Special Methods**: Such methods include, mutation breeding disruptive mating and selection and bi-parental mating. Biotechnology is expected to play significant role in in future in genetic improvement of Sorghum.

Breeding Centers

There are two types of *Sorghum* breeding centers as follows:

(i) **International Centers:** At International level *Sorghum* improvement work is carried out by International Crop Research Institute for Semi-arid Tropics (ICRISAT), Hyderabad (India). New varieties are developed after multi-location testing in different countries. Testing centres are located in Nigeria, Kenya, Zimbabve, Mexico and India.

(ii) **National Centers:** In India, *Sorghum* breeding work is carried out by following organizations:

(a) Indian Institute of Millet Research, Hyderabad. The new varieties are released through coordinated project after multi-location testing for 3-5 years.

(b) State Agricultural Universities located in *Sorghum* growing regions.

(c) Some private seed companies.

Practical Achievements

In India, several high yielding varieties and hybrids have been developed for different states. Hybrids are developed by using cytoplasmic genic male sterility. Kafir 60 is the source of male sterility. High lysine Ethiopian mutants (IS 11167 and 11758) are used for quality improvement. Some important high yielding varieties and hybrids of grain *Sorghum* are listed as follows:

(i) **Hybrids:** CSH 1, CSH 2, CSH 3, CSH 4, CSH 5, CSH 6, CSH 7, CSH 8, CSH 9, CSH 10 and CSH 11 for Kharif and, CSH 7, CSH 8, CSH 12, and CSH 13 for Rabi season.

(ii) **Varieties:** CSV 1, CSV 2, CSV 3, CSV 4, CSV 5, CSV 6, CSV 7, CSV 8, CSV 9, CSV 10 and CSV 11, CSV 12, CSV 13, M 35-1 and Swati for Kharif and, CSV 7 and CSV 8, for Rabi season.

Nucleus Seed Production

Nucleus seed is the base material in seed production chain. The important steps involved in the nucleus seed production of *Sorghum* are briefly presented as follows:

(i) **Base Material**: Nucleus seed is produced from nucleus seed or breeder seed or sometimes from foundation seed. Select good quality nucleus or breeder or foundation seed for production of nucleus seed. Grow the breeder seed in large plot of at least 0.25 acre. The field of breeder seed production already planted can be used for production of nucleus seed. There is no need of planting separate field for such purpose.

(ii) **Breeding Method Used**: In *Sorghum*, panicle to row selection is used for production of nucleus seed. One panicle true to the type is selected per plant at panicle emergence and subjected to selfing. Seed of each selected panicle is harvested separately and stored in separate envelope.

(iii) **Number of Panicle to be selected**: It depends on the quantity of nucleus seed to be produced. Generally, 200 to 500 panicles true to type of the variety are selected and seed of each selected panicle is harvested separately at maturity.

(iv) **Planting Progeny Rows**: In the second crop season progeny row of each selected panicle is planted separately. An isolation distance of 200m is kept all around the progeny rows. Each panicle to row progeny is carefully examined from panicle emergence onwards. Off type plants are removed from each progeny row. If variants in a progeny row are many, the entire progeny row is rejected.

(v) **Laboratory Evaluation:** In the laboratory, seed of all progeny of single plants are evaluated in terms of shape, size, colour and lustre.

(vi) **Seed Bulking:** Finally the seed of 200 true to the type progeny rows is bulked. This constitutes nucleus seed, which is used for production of breeder seed or nucleus seed in the third crop season. The bulk seed is grown in larger plots to produce breeder seed for further use in seed production chain.

TABLE 12.2: Steps in Nucleus Seed Production of Sorghum

Sl.No.	*Season*	*Particulars*	*Details*	*Remarks*
1	First	Base Population to be raised	0.25 Acre	In disease free field
		Panicles to be selected and selfed Panicles finally to be selected	1000 200	Tag selected ears
2	Second	Plant panicle to rows progeny	200	In isolation of 200m
		Laboratory evaluation of Ear row progeny	200	For seed characters
		Bulking equal number of seed from panicles	150	This constitutes nucleus seed
3	Third	Breeder seed production		Using nucleus seed

Note: Keep some nucleus seed in reserve for use in case of crop failure.

3. Pearl Millet

General Information

It includes, family, global distribution, national distribution, types, uses as briefly presented below:

(i) **Family**: Pearl millet belongs to the family *Poaceae* (old *Gramineae*). It is an important food crop of semi-arid tropics.

(ii) **Global Distribution**: Major pearl millet growing countries are Africa, India, Pakistan and South East Asia. In USA and Europe, pearl millet is grown as fodder crop.

(iii) National Distribution: In India, major pearl millet growing states are Rajasthan, Gujarat, Maharashtra, Uttar Pradesh and Haryana.

(iv) Uses: Grains are used as food and stalk as cattle fodder.

Cultivated Species

There is only one cultivated species of pearl millet, *i.e. Pennisetum americanum*. Earlier it was known as *P. typhoides*. The cultivated species is diploid ($2n = 14$). There are over 50 wild species of pearl millet in which chromosome number is multiple of 5, 7, 8 and 9. Two wild species, *viz.* Napier grass (*P. purpureum*) and *P. squamulatum* have been used in breeding programs to develop perennial fodder varieties, and drought and cold resistant cultivars, respectively.

Origin and Evolution

Vast genetic diversity in pearl millet is found in West Africa and, therefore, it is generally accepted that Western Africa is the original home of pearl millet. It spread to other countries from Africa in course of time. Nothing is so far known about the ancestor of cultivated pearl millet.

Reproduction and Pollination

(i) Pollination: Pearl millet is a cross-pollinated crop. In this crop natural cross pollination is more than 80 per cent. Pollen is very light and cross pollination occurs by wind.

(ii) Protogyny: It is common which promotes cross pollination.

(iii) Propagation: Pearl millet is a seed propagated crop. Some wild species of pearl millet such as Napier grass propagates asexually by means of rghizomes.

(iv) Tillering: In pearl millet, tillering is important.

(v) Isolation Distance: An isolation distance of 400 m. and 200 m is essential for the production of foundation and certified seed.

Breeding Objectives

In pearl millet, major breeding objectives are: higher grain yield, better quality, early maturity and wider adaptability and resistance to biotic (diseases and insects) and abiotic (drought and lodging) stresses as follows:

(i) Higher grain yield: Major Yield Components are: ear size and compactness, number of productive tillers per plant and test weight.

(ii) Grain quality: Bold, lustrous and pearly amber grains are considered desirable for human consumption in the form of various preparations.

(iii) Disease Resistance: Major diseases are Downy mildew, ergot and rust.

(iv) Insect Resistance: In India, Insects are not serious problems in this crop.

Breeding Procedures

Breeding approaches which are used for pearl millet improvement are of two types as presented below:

(i) General Methods: These methods include introduction, mass selection, back cross method, heterosis breeding, synthetic and recurrent selection. disruptive mating and selection, bi-parental mating.

(ii) Special Methods: Such methods include: tissue culture and biotechnology which are expected to play significant role in in future in genetic improvement of pearl millet.

Breeding Centers

There are two types of pearl millet breeding centers as follows:

(i) International Centers: At International level pearl millet improvement work is carried out by International Crop Research Institute for Semi-arid Tropics (ICRISAT), Hyderabad (India). New varieties are developed after multi-location testing in different countries. Testing centres are located in Nigeria, Kenya, Zimbabve, Mexico and India. ICRISAT maintains global gene pool of pearl millet. ICRISAT deals with five crops *viz.* pearl millet, *Sorghum*, chickpea, pigeon pea and groundnut.

(ii) National Centers: At National level, pearl millet breeding work is carried out by following organizations:

(a) Central Arid-zone Research Institute, Jodhpur: The new varieties are released through coordinated project after multi-location testing for 3-5 years. All India coordinate pearl millet improvement project is located at Agricultural Research Station, RAU, Mandore, Jodhpur (Rajasthan).

(b) State Agricultural Universities located in pearl millet growing regions.

(c) Some private seed companies.

Practical Achievements

Several high yielding varieties and hybrids of pearl millet have been developed in India for different pearl millet growing states. Two varieties, *viz.* ICMS 7703 and ICTP 8203 and two hybrids, *viz.* ICMH 423 and ICMH 451 have been released by ICRISAT for India. Important varieties developed by Research Stations of Agricultural Universities and other organizations are listed below:

(i) Hybrids: HHB 45, HHB 50, HHB 60, HHB 67 from Hisar; GHB 27, GHB 30, GHB 181 FROM Jamnagar [Gujarat]; Pusa 3 from IARI, New Delhi; PHB 17 from Punjab; Other hybrids include HBH 128, MBH 130, MH 143, MH 169, MH 179, MH 180, MH 82 MH 208, *etc.* All hybrids are CMS based. Tift 23 A and Ludhiana cytoplasm have been used as sources of CMS in pearl millet improvement.

(ii) Varieties: PSB 8, PSB 15, PSB 2, HC 4, RHR 1, PCB 15, Mukta, Mallikarjun, Pusa safed Bajra,

Nucleus Seed Production

Nucleus seed is the base material in seed production chain. The important steps involved in the nucleus seed production of pearl millet are briefly presented as follows:

(i) **Base Material:** Nucleus seed is produced from nucleus seed or breeder seed or sometimes from foundation seed. Select good quality nucleus or breeder or foundation seed for production of nucleus seed. Grow the breeder seed in large plot of at least 0.25 acre.

(ii) **Breeding Method Used:** In pearl millet, Ear to row selection is used for production of nucleus seed. One Ear true to the type is selected per plant at the time of ear emergence. At maturity, seed of each selected Ear is harvested separately and stored in separate envelope. The field of breeder seed production already planted can be used for production of nucleus seed. There is no need of planting separate field for such purpose.

(iii) **Number of Ear to be selected:** It depends on the quantity of nucleus seed to be produced. Generally, 500 to 1000 Ears true to type of the variety are selected, selfed and seed of each selected Ear is harvested separately.

(iv) **Planting Progeny Rows:** In the second crop season progeny row of each selected Ear is planted separately. An isolation distance of 1000m is maintained all around the progeny rows. Each Ear to row progeny is carefully examined from ear emergence onwards and off type plants are removed from each progeny row. If variants in a progeny row are many, the entire progeny row is rejected.

(v) **Laboratory Evaluation:** In the laboratory, seed of all progeny of single plants are evaluated in terms of shape, size, colour and lustre.

(vi) **Seed Bulking:** Finally the seed of 200 true to the type progeny rows is bulked. This constitutes nucleus seed, which is used for production of breeder seed or nucleus seed in the third crop season. The bulk seed is grown in larger plots to produce breeder seed for further use in seed production chain.

TABLE 12.3: Steps in Nucleus Seed Production in Pearl Millet

Sl.No.	*Season*	*Particulars*	*Details*	*Remarks*
1	First	Base Population to be raised	0.25 Acre	Self all plants
		Single plants to be selected	200	Selfed Plants
2	Second	Progeny rows of single plants	200	In isolation
		Laboratory evaluation of single plant progeny	100	For seed characters
		Mixing true to type progeny rows	75	This constitutes nucleus seed
3	Third	Breeder seed production		Using nucleus seed

Note: Keep some nucleus seed in reserve for use in case of crop failure.

Summary

Important millet crops include maize, Sorghum and pearl millet. An area of plant breeding which deals with principles and methods of breeder seed production is referred to as maintenance breeding or varietal maintenance technology. Main points related to base material, breeding procedure used, main focus, plant part used, progenies grown, isolation distance, selection of field, crop inspection and evaluation of selected progenies for maintenance breeding [nucleus seed production] of maize, Sorghum and pearl millet crops have been discussed.

QUESTIONS

1. **Describe briefly genetic origin, species composition and breeding objectives of maize**
2. **Discuss in brief breeding procedure used for genetic improvement of maize crop.**
3. **Explain in short practical achievements of maize breeding in India.**
4. **Describe in brief important maize breeding centres and their role in varietal improvement.**
5. **Describe briefly genetic origin, species composition and breeding objectives of Sorghum.**
6. **Discuss in brief breeding procedure used for genetic improvement of Sorghum crop.**
7. **Explain in short practical achievements of Sorghum breeding in India.**
8. **Describe in brief important Sorghum breeding centres and their role in varietal improvement.**
9. **Describe briefly genetic origin, species composition and breeding objectives of pearl millet in India.**
10. **Discuss in brief breeding procedure used for genetic improvement of pearl millet.**
11. **Explain in short practical achievements of pearl millet breeding in India.**
12. **Describe briefly the procedure of nucleus seed production in maize.**
13. **Explain briefly the procedure of nucleus seed production in Sorghum.**
14. **Discuss briefly the procedure of nucleus seed production in pearl millet.**

* * * *

Chapter 13

Nucleus Seed Production of Commercial Crops

Introduction

Important commercial field crops include sugarcane, tobacco and potato. A brief account of distribution, cultivated species, origin and evolution, reproduction and pollination, breeding objectives, breeding procedures, breeding centers, practical achievements and nucleus seed production related to these crops is presented in this chapter.

1. Sugarcane (*Saccharum* sp.)

General Information

(i) **Family:** Sugarcane belongs to the family *Gramineae*. It is an important commercial crop which is grown in tropical and subtropical climates the world over. Being C4 plant, it has high photosynthetic efficiency.

(ii) **Global Distribution**: The important sugarcane growing countries are India, Brazil, Cuba, China, USA, Mexico, France, Germany and Australia.

(iii) **National Distribution**: In India, the major sugarcane growing states are Uttar Pradesh, Maharashtra, Andhra Pradesh, Tamil Nadu, Karnataka, Bihar, Punjab and Haryana.

(iv) **Uses**: Sucrose is the main product of sugarcane. The other products are green tops, bagasse, molasses, filter mud, stubbles, and dry leaves. Green tops are used as cattle feed; bagasse for production of paper, rayon grade pulp, particle boards, fibre boards, as a fuel and also for production of alcohol; molasses for production of citric acid, butamol, acetone, rum, glycol, alcohol and acetic acid; filter mud for production of crude wax, fatty lipids, waxy lipids and refined wax; stubles for making compost and as fuel; and dry leaves as mulch, fuel and compost.

Cultivated and Wild Species

There are three cultivated species of sugarcane, *viz. Saccharum officinarum* ($2n$ = 80), *S. sinense* ($2n$ = 118) and *S. barberi* ($2n$ = 82-124). The species *S.officinarum* is also known as noble cane. There are two wild species of sugarcane, *viz. S. spontaneum* and *S. robustum*. Main features of wild and cultivated species are as follows:

TABLE 13.1: Main Features of Wild and Cultivated Species of Sugarcane

Sl.No.	*Species*	*Distribution*	*Main Characters*
1	*Saccharum officinarum*	Grown in South India	Canes are thick, susceptible to drought, red rot and mosaic. Soft rind, low fibre, high sugar content, *etc.*
2	*S. barberi*	Grown in North India	Canes are thin with hard rind, early maturing, wider adaptability, low sugar content and resistance to red rot butsusceptible to smut.
3	*S. sinense*	Grown in North India	Canes are thin with hard rind, early maturing, wider adaptability, low sugar content and resistance to red rot but susceptible to smut.
4	*S. spontaneum*	Found in India, South and South East Asia,	Canes are thin with hard rind. This has high tillering capacity and wider adaptability. early maturing, low sugar content and resistance to red rot but susceptible to smut.
5	*S. robustum*	Found in New Guinea	Wider adaptability, very thick canes, low sugar content, high fibre and susceptible to mosaic.

Origin and Evolution

It is believed that *Saccharum officinarum* has originated in New Guinea and *S. barberi* and *S. sinensis* in North India. Wild species *S. robustum* is considered to be the progenitor of cultivated species *S. officinarum*. In other words, noble cane originated in New Guinea and migrated to North India. After hybridization with wild species *S. spontaneum*, it gave birth to *S. barberi* and *S. sinense*. The overall process is presented as follows:

1. *S. robustum* ————→ *S. officinarum* -→ New Guinea
2. *S. offinarum* x *S. spontaneum* —→ *S. barberi* and *S. sinense* —→ North India.

Reproduction and Pollination

Sugarcane is a cross-pollinated and asexually propagated crop. Main points are presented as follows:

(i) **Flowers**: Flowers are bisexual and inflorescence in called as arrow. Flowering requires warm nights, humid conditions and high rain fall. These conditions are available at Coimbatore.

(ii) **Pollination**: Sugarcane is a cross-pollinated crop.

(iii) **Apomixis**: In sugarcane apomixis is not reported.

(iv) **Propagation**: Sugarcane is asexually propagated.

(v) **Tillering**: Tillering is important in this crop.

(vi) Isolation Distance: An isolation distance of 50 m and 30 m is required for production of foundation of and certified seed.

Breeding Objectives

In sugarcane, major breeding objectives are higher cane yield, better fibre quality, early maturity, wider adaptability and resistance to biotic and abiotic stresses as follows:

(i) Higher cane yield: Major yield components are: stem height, thickness, tillering capacity, number of millable canes per plant and weight of individual cane.

(ii) Quality: Milling quality includes soft rind, long internode, less fibre, less pith, high sucrose content and thick juice.

(iii) Disease Resistance: Major diseases include red root and smut.

(iv) Insect Resistance: Important insects include top borer, pyrilla, white fly, early shoot borer, and internode borer.

Breeding Procedures

Breeding methods which are used for the genetic improvement of sugarcane can be divided into three groups, *viz.* general methods, special methods and population improvement procedures as follows:.

(i) General Methods: These methods include introduction, clonal selection and backcrossing.

(ii) Special Methods: These methods include interspecific hybridization, inter-generic hybridization and mutation breeding. Distant hybridization (inter specific and inter-generic crosses) is used for transfer of resistance to biotic and abiotic stresses from wild species or allied genera into the cultivated species. Mutation breeding is mainly used for disease resistance.

(iii) Population Improvement: In sugarcane, population improvement program is also taken up. Bi-parental mating, disruptive mating and recurrent selection are used for population improvement. Tissue culture technique is used for development of haploids (anther-culture), earliness and for resistance to biotic and abiotic factors. Somaclonal variation is expected to help in evolving varieties resistant to biotic and abiotic stresses.

Breeding Centers

Sugarcane breeding work is carried out in each country at national research centre. No International Research Institute has been established for sugarcane breeding so far. In India, sugarcane breeding work is carried out by the following research Institutes/organizations:

1. Sugarcane Breeding Institute, Coimbatore.
2. Indian Institute of Sugarcane Research, Lucknow.

3. State Sugarcane Research Stations, such as Shahjahanpur (UP), Seorahi, Deoria (UP), Pusa (Bihar), Padegaon, (Maharashtra) and Anakapalli (AP). New varieties are released by All India coordinated Sugarcane Improvement Project after multi-location testing for 3-5 years. Coordinated Project is located at Lucknow in the premises of ISSR.

Practical Achievements

Several improved varieties of sugarcane suitable for different states have been released by Sugarcane Breeding Institute, Coimbatore and State Sugarcane Research Stations through Coordinated Project. Some currently cultivated sugarcane varieties are Co 1148, Co 1158, Co 64, Co 997, Co 6304, Co 419, Co 740, Bo 17 and Co Pant 84211. Some varieties with special characters are given below:

1. **Early varieties**: Co 8336, Co 8337, Co 8338, Co 8339, Co 8340, Co 8341, (8-9 months) Co J64, and Co C 671.
2. **Resistant to abiotic stresses**

   Drought: Co 285, Co 740, Co 997, Co 1148.

   Frost: Co 1148, N Co 310.

   Salinity: Co 453, Co 62125.

   Lodging: Co 6304, Co 7117, CoS 7918.

   Water logging: Co 1157, Co 975, Co 785, Bo 91, Bo 104, Bo 106, Bo 109, Bo 120, Cos 510, CoS 767, CoS 8016.
3. **Resistant to biotic stresses**

   Top borer: COJ 67, Co 1158, Co 7224.

   Internode borer: COC 671, Co 975, Co 6806, Co 62175.

   Red rot: Co 7627, COJ 64, COR 8001; CoLK 8001, COH 7803, Bo 108.

Nucleus Seed Production

Nucleus seed is the base material in seed production chain. The important steps involved in the nucleus seed production of sugarcane are briefly presented as follows:

(i) **Base Material**: Nucleus seed is produced from nucleus seed or breeder seed or sometimes from foundation seed. Select good quality nucleus or breeder or foundation seed for production of nucleus seed. Grow the breeder seed in large plot of at least 0.25 acre. The field of breeder seed production already planted can be used for production of nucleus seed. There is no need of planting separate field for such purpose.

(ii) **Breeding Method Used**: In sugarcane clump to row selection is used for production of nucleus seed. Single clump true to the type is selected per plant during tillering stage. Seed of each selected clump is marked, numbered and harvested.

(iii) **Number of Clump to be selected**: It depends on the quantity of nucleus seed to be produced. Generally, 100 clumps true to type of the variety are randomly selected from the base population and each selected clump is harvested separately.

(iv) **Planting Progeny Rows**: In the second crop season progeny row of each selected clump is planted separately adopting proper isolation distance. Each clump to row progeny is carefully examined from tillering onwards. Off type plants are removed from each progeny row. If variants in a progeny row are many, the entire progeny row is rejected.

(v) **Bulking Clumps:** The seed of all true to the type progeny rows is bulked. This constitutes nucleus seed, which is used for production of breeder seed or nucleus seed in the third crop season. The bulk seed is grown in larger plots to produce breeder seed for further use in seed production chain.

TABLE: 13.2: Steps in Nucleus Seed Production of Sugarcane

Sl.No.	*Season*	*Particulars*	*Details*	*Remarks*
1	First	Base Population to be raised	5000 Hills	In disease free field
		Single clumps to be selected	200	Tag selected ears
2	Second	Plant clump to rows progeny	200	In isolation of 5m
		Laboratory evaluation of progeny of clumps	150	For seed characters
		Bulking equal number of seed from best clumps.	100	This constitutes nucleus seed
3	Third	Breeder seed production		Using nucleus seed

2. Tobacco

General Information

(i) **Family:** Tobacco belongs to the family *Solanaceae*. It is a narcotic plant which has commercial value. It is grown in tropical and subtropical climates the world over.

(ii) **Global Distribution**: The important tobacco growing countries are USA, China, India, Thailand, Pakistan, Myanmar, Sri Lanka, Indonesia, *etc.*

(iii) **National Distribution**: In India, the major tobacco growing states are Andhra Pradesh, Gujarat, Karnataka, Maharashtra, Tamil Nadu, Uttar Pradesh, Bihar and West Bengal.

(iv) **Types**: Cultivated tobacco is of two types, *viz. Nicotiana rustica* and *Nicotiana tabaccum*. Depending on use, tobacco is classified into various groups such as cigarette (flue cured) tobacco, Bidi tobacco, Hookah tobacco, Cigar and Cheroot type, chewing and snuff type, *etc.*

(v) **Uses**: Tobacco is used for chewing, smoking and medicinal uses. Smoking is done through cigarette, Bidi, Hookah, Cigar and Cheroot.

Cultivated Species

There are two cultivated species of tobacco, *viz. Nicotiana rustica* and *Nicotiana tabaccum*. Both species are tetraploid [2n=4x=48]. *Nicotiana rustica* requires cooler climate whereas *N. tabaccum* can be grown throughout India. *Nicotiana tabaccum* is the predominant species which accounts for 90 per cent of the total tobacco area. The remaining 10 per cent is covered by *N. rustica*. There are more than 60 wild species of tobacco which are found in South and North America, Australia and South Pacific Islands. Wild species are important sources of resistance to biotic and abiotic stresses.

Origin and Evolution

Andes region is the original home of tobacco. There are two cultivated species of tobacco [*Nicotiana tabacum* and *N. rustica*]. Both these cultivated species have developed from a cross between two different wild diploid species as follows:

(*a*) *N. sylvestris* ($2n$ = 24) x N. *tomentosa* ($2n$ = 24) = N. tabacum = (2n= 48)

(*b*) *N. paniculata* x *N. undulata.* = *N. rustica* = (2n =48)

(i) Origin of Nicotiana tabacum

Parents	*Nicotiana sylvestris*	x *Nicotian tomentosa*
Chromosome	[2n=24]	[2n=24]
F_1	[2n=24] Sterile	
Chromosome doubling in nature	[2n =48] *Nicotiana tabacum* [Tetraploid-Fertile]	

(ii) Origin of Nicotiana rustica

Parents	*Nicotiana paniculata*	x *Nicotian undulata*
Chromosome	[2n=24]	[2n=24]
F_1	[2n=24] Sterile	
Chromosome doubling in nature	[2n =48] *Nicotiana rustica* [Tetraploid-Fertile]	

Figure 13.1: Probable Origin of Cultivated Tetraploid Species of Tobacco.

Clausen and Goodspeed [1925] synthesized a new hexaploid species of tobacco (*Nicotiana*) from a cross between *Nicotiana tabacum* ($2n$ - 48) and *N. glutinosa* ($2n$ = 24). The F_1 was sterile with $2n$ = 36, which was made fertile by doubling of chromosome through colchicine treatment The new species, is known as *N. digluta*.

Parents	*Nicotiana tabacum*	x *Nicotian sylvestris*
Chromosome	[2n=48]	[2n=24]
F_1	[2n=36] Sterile	
Chromosome doubling	[2n =72] *Nicotiana digluta* [Hexaploid-Fertile]	

FIGURE 13.2: Probable Origin of Hexaploid Species of Tobacco.

Reproduction and Pollination

Tobacco is a self-pollinated and seed propagated crop. Main points are presented as follows:

(i) **Flowers**: Flowers are bisexual and showy. Flowering requires warm nights and humid conditions.

(ii) **Pollination**: Tobacco is a self-pollinated crop. Outcrossing is less than 5 per cent.

(iii) **Apomixis**: In tobacco apomixis is not reported.

(iv) **Propagation**: Tobacco is a seed propagated crop.

(v) **Isolation Distance**: An isolation distance of 5 m is required for the production of foundation and certified seed.

Breeding Objectives

In tobacco, major breeding objectives are higher yield, acceptable quality and resistance to diseases and insects as follows:

(i) **Higher Leaf yield**: Major yield components are number of leaf per plant, leaf size and weight of single leaf.

(ii) **Acceptable quality**. Quality characters include nicotine content, sugar content and aroma. High nicotine is preferred in Bidi, hookah and chewing tobacco and low in cigarette tobacco. High sugar content is preferred in cigarette tobacco. In USA, cigarette tobacco contains about 18 per cent sugar. Thin leaves are preferred for cigar and pipe smoking, thicker leaves for cigarettes and the thickest leaves for chewing.

(iii) **Disease Resistance**: Major diseases black shank, bacterial blight, Fusarium wilt, powdery mildew, anthracnose, mosaic leaf curl, *etc.*

(iv) **Insect Resistance**: The main insect is tobacco bud worm.

Breeding Procedures

The main breeding methods used in tobacco are of two types, *viz.* general breeding methods and special breeding methods as follows:

(i) **General Methods**: General methods include, introduction, pure line selection, mass selection, backcross, pedigree selection and single seed descent method.

(ii) **Special Methods**: Such methods include mutation breeding, distant hybridization and biotechnology. Mutation breeding is used especially for creating variability for various economic characters. Distant hybridization is used for interspecific gene transfer. Biotchnology especially tissue culture technique is useful in developing haploids, and interspecific gene transfer adopting embryo rescue and protoplast fusion techniques.

Breeding Centers

In India, tobacco improvement work is carried out by following organizations:

(i) Central Tobacco Research Institute, (CTRI) Rajahmundry, Andhra Pradesh,

(ii) All India Coordinated Tobacco Improvement Project (AICTIP), Anand (Gujarat) and

(iii) State Agricultural Universities in tobacco growing states: New varieties are released through coordinated project after multi-locational testing for 3-5 years.

Practical Achievements

In India, several high yielding and good quality varieties of tobacco suitable for various uses have been released for different states through CTRI, AICTIP and SAUs. Some recently released varieties are listed below:

1. **Flue cured varieties**: CTRI spl, Jayasri, Godavari spl, Swarna, Hema, Bhavya, *etc.*
2. **Cheroot**: Bhavani spl, Lanka spl, Viswanadh, Prabhat, *etc.*
3. **Chewing**: Thangam, Bhagyalakshmi, Maragandham, Sona, *etc.*
4. **Cigar filters**: Krishna.

Nucleus Seed Production

Nucleus seed is the base material in seed production chain. The important steps involved in the nucleus seed production of tobacco are briefly presented as follows:

(i) Base Material: Nucleus seed is produced from nucleus seed or breeder seed or sometimes from foundation seed. Select good quality nucleus or breeder or foundation seed for production of nucleus seed. Grow the breeder seed in in nursery of 1m wide and 10m length to raise seedlings for transplanting. The field of breeder seed production already planted can be used for production of nucleus seed. There is no need of planting separate field for such purpose.

(ii) Breeding Method Used: In tobacco, single capsule may yield 2000 to 10,000 seeds. Hence, a single plant is selected from the base population for production of nucleus seed. One capsule from true to the type plant is selected at maturity. Seed of selected capsule is harvested separately and stored in separate envelope.

(iii) Number of Capsule to be selected: It depends on the quantity of nucleus seed to be produced. Generally, one capsule from plant true to type of the variety is selected and seed of selected capsule is harvested separately.

(iv) Planting Progeny Rows: In the second crop season, nursery of selected capsule is raised in nursery of 10m x 1m size. These seedlings are planted in the field and the progeny plants are carefully examined from flowering onwards. Off type plants are removed from each progeny row. If variants in a progeny row are many, the entire progeny row is rejected.

(v) **Bulking Seed:** The seed of all true to the type progeny rows is bulked. This constitutes nucleus seed, which is used for production of breeder seed or nucleus seed in the third crop season. The bulk seed is grown in larger plots to produce breeder seed for further use in seed production chain.

3. Potato

General Information

(i) **Family:** Potato belongs to the family *Solanaceae*. It is an important vegetable crop which is grown in temperate regions the world over.

(ii) **Global Distribution**: Europe and Asia together account for 85 per cent of world potato production. Main potato producing countries are USA, Russia, India, China, *etc.*

(iii) **National Distribution**: In India, the major potato growing states are Himacha Pradesh, Punjab, Uttar Pradesh and Bihar.

(v) **Uses**: Potato is used as important vegetable crop. Hence, it is called as king of vegetables. Several products are made from potato.

Cultivated Species

There are four types of cultivated potato, *viz.* diploid [2n=24], triploid [3x =36], tetraploid [4x = 48] and pentaploid [5x =60]. The most important cultivated species is *Solanum tuberosum* which is grown throughout the World. *S. tuberosum* is cultivated mainly in high hills of South America. Diploid, triploid and pentaploid species are cultivated in Peru and Bolivia. Several wild species of potato are found in USA and Mexico.

Origin and Evolution

The Central Andean region of South America is the original home of potato. Cultivated potato [*Solanum tuberosum*] is a tetraploid species [2n=48] which is believed to have originated from the diploid species *S. stenotomum* which is also found growing in the Andes region of South Peru and Bolivia. Eralier it was believed that tetraploid cultivated species *S. tuberosum* has originated through chromosome doubling from wild diploid species *S. stenotomum* found in the Andes region of South Peru and Bolivia.According to recent view, it is probably a segmental allopolyploid derived from the cross between cultivated diploid species *S. stenotomum* and wild diploid species *S. sparsipillum* [Hawkes, 1956, 58].

Reproduction and Pollination

Potato is a self-pollinated and vegetatively propagated crop. Main points are presented as follows:

(i) Flowers: Flowers are bisexual.

(ii) Pollination: Potato is a self-pollinated crop.

(i) Earlier view

Solanum stenotomum **→ chromosome doubling →** ***Solanum tuberosum***

(ii) Recent View

Parents	*Solanum stenotomum*	x *S. sparsipillum*
Chromosome	[2n=24]	[2n=24]
F_1	[2n=24] sterile	
Chromosome doubling in nature	[4n =48] *Solanum tuberosum* -fertile	

FIGURE 13.3: Probable Origin of Cultivated Tetraploid Potato.

(iii) Self-incompatibility: In potato self- incompatibility is reported.

(iv) Propagation: Potato propagates by tubers.

Breeding Objectives

In potato, major breeding objectives are higher yield, quality, earliness, wider adaptation, resistance to diseases, insects, herbicides, heat, drought and frost as follows:

(i) **Higher Tuber yield**: Major yield components are number of tubers per plant and tuber weight.

(ii) **Acceptable quality:** Quality characters include shape, size, color, skin thickness, position of eyes and nutritive value. Tubers with external (cracks, knobs) and internal (hollow heart, necrosis and pigmentation) defects are discarded. High solid contents, good flavour and texture and low glycol-alkaloid contents come under quality characters.

(iii) **Disease Resistance**: Major diseases are late blight, potato virus x and y, charcoal rot, potato wart and early blight.

(iv) **Insect Resistance**: The main insects are aphids, beetle and nematodes.

Breeding Procedures

Breeding methods which are used in potato are of two types, *viz.* general methods and special methods as follows:

(i) **General Methods**: Such methods include introduction, clonal selection, backcrossing and recurrent selection.

(ii) **Special Methods:** Such methods include interspecific hybridization and induced mutations. In potato, superior genotype can be selected at any stage and can be maintained easily by vegetative reproduction. There is no need for a selected individual to breed true. Asexual propagation permits fixation of heterosis. Recurrent selection is used for improvement of parental populations.

Biotechnology is expected to play an important role in future in potato improvement. Somaclonal variation will be useful in developing varieties resistant to biotic and abiotic stresses including herbicides. Biotechnology will also help in interspecific gene transfer.

Breeding Centers

There are two types of potato breeding centers as follows:

(i) **International Centers:** International potato Centre (CIP), Peru serves as a world centre for genetic improvement of potato. The global collection of potato is also maintained at this centre. The main focus of this centre is on the improvement of potato for tropical and temperate climates of different countries. It has testing centres in different countries.

(ii) **National Centers:** In India, potato breeding work is carried out by following organizations:

   (a) **Central Potato Research Institute (CPRI), Shimla (Himachal Pradesh).** Potato requires long day conditions, and monsoon climate for flowering. These conditions are available in Shimla. This is why potato Research Institute has been located in Shimla. CPRI has its regional stations in Ooty, Darjeeling and Mukteshwar (Nainital) in the hill regions, besides centres in plains at Jalandhar, Meerut, Patna, Bangalore, Gwalior. The seed multiplication work is carried out at Modipuram (Meerut).

   (b) **State Agricultural Universities:** The new varieties are released by All India Coordinated Potato Improvement Project (AICPIP), Shimla through multi-location testing for 3-5 years.

Practical Achievements

CPRI, Shimla has done remarkable work on potato improvement. Several high yielding varieties for various areas have been developed by this Institute. Some important varieties are Kufri Red, Kufri Sindhuri, Kufri Chamatkar, Kufri Kuber, Kufri Kundan, Kufri Alankar, Kufri Chandramukhi and Kufri Jyoti.

Nucleus Seed Production

Nucleus seed is the base material in seed production chain. The important steps involved in the nucleus seed production of potato are briefly presented as follows:

(i) **Base Material**: Nucleus seed is produced from nucleus seed or breeder seed or sometimes from foundation seed. Select good quality nucleus or breeder or foundation seed for production of nucleus seed. Grow the breeder seed in large plot of at least 0.25 acre. The field of breeder seed production already planted can be used for production of nucleus seed. There is no need of planting separate field for such purpose.

(ii) **Breeding Method Used**: In potato, plant to progeny row selection is used for production of nucleus seed. Single plants true to the type are randomly selected from the field of base population at maturity. Tubers of each selected plant are harvested separately and stored separately.

(iii) **Number of Plants to be selected**: It depends on the quantity of nucleus seed to be produced. Generally, 100 to 150 single plants true to type of the variety are selected, and tubers of each selected plant are harvested separately.

(iv) **Planting Progeny Rows**: In the second crop season progeny row of each selected plant is planted separately. Each plant to row progeny is carefully examined at grand growth period. Off type plants are removed from each progeny row. If variants in a progeny row are many, the entire progeny row is rejected.

(v) **Bulking Seed:** The tubers of all true to the type progeny rows are bulked. This constitutes nucleus seed, which is used for production of breeder seed or nucleus seed in the third crop season. The bulk seed is grown in larger plots to produce breeder seed for further use in seed production chain.

Summary

Important commercial field crops include sugarcane, tobacco and potato. An area of plant breeding which deals with principles and methods of breeder seed production is referred to as maintenance breeding or varietal maintenance technology. Main points related to base material, breeding procedure used, main focus, plant part used, progenies grown, isolation distance, selection of field, crop inspection and evaluation of selected progenies for maintenance breeding [nucleus seed production] of sugarcane, tobacco and potato are discussed.

QUESTIONS

1. **Describe briefly genetic origin, species composition and breeding objectives of Sugarcane.**
2. **Discuss in brief breeding procedure used for genetic improvement of Sugarcane crop.**
3. **Explain in short practical achievements of Sugarcane breeding in India.**
4. **Describe in brief important Sugarcane breeding centres and their role in varietal improvement.**
5. **Describe briefly genetic origin, species composition and breeding objectives of Potato in India.**
6. **Discuss in brief breeding procedure used for genetic improvement of potato crop.**
7. **Explain in short practical achievements of potato breeding in India.**

8. **Describe briefly genetic origin, species composition and breeding objectives of Tobacco in India.**
9. **Discuss in brief breeding procedure used for genetic improvement of Tobacco crop.**
10. **Explain in short practical achievements of Tobacco breeding in India.**
11. **Describe briefly the procedure of nucleus seed production in sugarcane.**
12. **Explain briefly the procedure of nucleus seed production in Tobacco.**
13. **Discuss briefly the procedure of nucleus seed production in Potato.**

* * * *

Chapter 14

Nucleus Seed Production of Oilseeds

Introduction

There are several oil yielding crops such as rapeseed and mustard, ground nut, soybean, sunflower, *etc.* A brief account of distribution, cultivated species, origin and evolution, reproduction and pollination, breeding objectives, breeding procedures, breeding centers, practical achievements and nucleus seed production related to these crops is presented in this chapter.

1. Rapeseed and Mustard

General Information

(i) **Family**: Rapeseed and mustard are important oilseed crops which belong to the family *Cruciferae*. They contribute more than 13 per cent to the global production of edible oil. Rapeseed and mustard collectively are known as oilseed brassicas. *Brassicas* can be grown in cool season subtropics, higher elevations and as winter crops in the milder region of temperate zone.

(ii) **Global Distribution**: Major rapeseed and mustard growing countries are China, Canada, the Indian subcontinent and northern Europe. China, India and Pakistan collectively contribute more than 90 per cent to the global production.

(iii) **National Distribution**: In India, major *Brassica* producing states are, Uttar Pradesh, Rajasthan, Punjab, Assam, Bihar and West Bengal.

(iv) **Uses**: The seeds of *brassica* contain 40 to 45 per cent oil and 38 to 41 per cent protein. In India, oil is used for human consumption and cake as livestock feed.

Cultivated Species

There is one main cultivated species of rapeseed *i.e. Brassica campestris* (2n=20).

This species has three ecotypes, *viz.* yellow sarson (*Brassica campestris* var yellow sarson), brown sarson (*Brassica campestris* var brown sarson) and toria (*Brassica campestris* var toria). Yellow sarson and brown sarson are collectively called as turnip rape and Toria is known as Indian rape. The other cultivated species is *Brassica napus* (2n=38). In India, there are two main cultivated species of mustard, *viz.* Rai or Indian mustard (*Brassica juncea* 2n=4x=36) and Banarsi Rai or Black mustard (*Brassica nigra* 2n=16). There are 36 species in the genus Brassica. Most of the species are non oleiferous.

Origin and Evolution

It is believed that Rapeseed has (*Brassica campestris*) has originated in Himalayan region. Middle East is considered as primary centre of origin of Indian Mustard. China and North East India are considered as secondary centres of origin. *Brassica juncea* is amphidiploid between *B. nigra* and *B. campestris, Brassica carinata* between *B. nigra* and *B. oleracea*, and *Brassica napus* between *B. oleracea* and *B. campestris*.

In *brassica*, there are three basic species, *viz.* black mustard [*Brassica nigra*] (*BB,*

(i) Origin of *Brassica carinata*

Parents *Brassica nigra* x *Brassica oleracea*

Genome BB[2n=16] CC[2n=18]

F_1 BC [2n=17] sterile

Chromosome doubling in nature BBCC [2n =34] *B. caranata*- fertile

(ii) Origin of *Brassica napus*

Parents *Brassica campestris* x *Brassica oleracea*

Genome AA[2n=20] CC[2n=18]

F_1 AC [2n=19] sterile

Chromosome doubling in nature AACC [2n =38] *B. napus*- fertile

(iii) Origin of *Brassica juncea*

Parents *Brassica campestris* x *Brassica nigra*

Genome AA[2n=20] BB [2n=16]

F_1 AB [2n=18] sterile

Chromosome doubling in nature AABB[2n =36] *B. juncea* -fertile

FIGURE 14.1: Origin of Tetraploid Species of Brassica as per Nagharu U.

$2n = 16$), cabbage mustard [*B. oleracea* (*CC*, $2n = 18$)] and yellow sarson [*B. campestris* (*AA*, $2n = 20$)]. Three new tetraploid species have originated from a cross between two different diploid species as follows:

(*a*) *B. nigra* x *B. oleracea* = *B. carinata.*

(*b*) *B. compestris* x *B. oleracea* = *B. napus*

(*c*) *B. campestris* x *B. nigra* = B. *juncea.* [Indian mustard]

Reproduction and Pollination

Rapeseed and mustard are seed propagated crops. Main points are presented as follows:

(i) Pollination: *Brassica napus* and *Brassica juncea* are self-pollinated species with natural outcrossing up to 30 per cent. Indian yellow sarson is self-compatible.

(ii) Self-incompatibility: Brown sarson and Toria are self-incompatible. *Brassica nigra* and *B. oleracea* are also self-incompatible. Self-incompatibility promotes cross pollination in these species which occurs by wind and honey bees.

(iii) Propagation: Rapeseed and mustard are seed propagated crops.

(iv) Isolation Distance: An isolation distance of 400 meters for the production of foundation seed and 200 meters for certified seed is required under Indian conditions.

Breeding Objectives

Main breeding objectives in rapeseed and mustard are higher yield, early maturity, shattering resistance, ideal plant type, better quality and resistance to biotic (diseases and insects) and abiotic (drought and temperature) stresses.

(i) Higher grain yield: Major yield components include number of primary and secondary branches, number of siliquae per plant, seed per siliqua and seed size.

(ii) Quality Traits: Quality includes high seed oil content and low erucic acid and glucosinolate in the oil.

(iii) Disease Resistance: Major diseases include alternaria blight, white rust, downey mildew, *etc.*

(iv) Insect Resistance: Aphid is the major insect pest of rapeseed and mustard.

Breeding Procedures

There are two types of rapeseed and mustard species, *viz.* self-compatible and self-incompatible (cross pollinated).

(i) Self-Pollinated Species: In self-pollinated species, pure line selection, back crossing, pedigree method, single seed descent method and bulk breeding methods are commonly used.

(ii) Cross-Pollinated Species: In cross-pollinated species, mass selection, progeny selection, recurrent selection and synthetic breeding are generally used.

(iii) Distant hybridization is used for interspecific or intergeneric gene transfer.

(iv Natural polyploidy has resulted in the development of three amphidiploid species (*B. napus, B. juncea* and *B. carinata*). Mutation breeding is used for creating variability for various economic characters.

(v) Somaclonal variation has been recorded for earliness, dwarfness, high yield and resistance to diseases and drought (Chopra *et al.*, 1989). Haploids and transgenic plants have been developed through biotechnology.

Breeding Centers

In India, Rapeseed and mustard breeding work is carried out by National Research Centre for Mustard, Bharatpur (Rajasthan) and State Agricultural Universities of Rapeseed and Mustard growing states. New varieties are released after multi-location testing in coordinated trials for 3-5 years. The coordinated project was located at HAU, Hisar. Now this has been shifted to Bharatpur. Several high yielding varieties of rapeseed and mustard have been released in India for different states by above research centres.

Practical Achievements

Important National varieties of mustard are Kranti, RLM 198, Krishna and RLM 514. Pant Toria 303 is the national variety of rapeseed.

Nucleus Seed Production

Nucleus seed is the base material in seed production chain. The important steps involved in the nucleus seed production of Rapeseed and Mustard are briefly presented as follows:

(i) Base Material: Nucleus seed is produced from nucleus seed or breeder seed or sometimes from foundation seed. Select good quality nucleus or breeder or foundation seed for production of nucleus seed. Grow the breeder seed in large plot of at least 0.25 acre. The field of breeder seed production already planted can be used for production of nucleus seed. There is no need of planting separate field for such purpose.

(ii) Breeding Method Used: In rapeseed and mustard, plant to progeny selection is used for production of nucleus seed. Plants true to the type are selected during flowering in self-pollinated species. Seed of each selected plant is harvested separately and stored in separate envelope.

(iii) Number of Plants to be selected: It depends on the quantity of nucleus seed to be produced and mode of pollination of the crop species. Generally, 500 plants true to type of the variety in self-pollinated species and 2500 plants in cross-pollinates species [self-incompatible species] are selected, and seed of each selected plant is harvested separately.

(iv) Planting Progeny Rows: In the second crop season progeny row of each selected plant is planted separately adopting proper isolation distance. Each plant to row progeny is carefully examined from flowering onwards. Off type plants are removed from each progeny row. If variants in a progeny row are many, the entire progeny row is rejected.

(v) **Seed Bulking:** The seed of all true to the type progeny rows is bulked. This constitutes nucleus seed, which is used for production of breeder seed or nucleus seed in the third crop season. The bulk seed is grown in larger plots to produce breeder seed for further use in seed production chain.

TABLE 14.1: Steps in Nucleus Seed Production of Rapeseed and Mustard

Sl.No.	*Season*	*Particulars*	*Details*	*Remarks*
1	First	Base Population to be raised	0.25 Acre	In disease free field
		Single plants to be selected		Tag selected plants
		Self-pollinated	500	
		Cross-pollinated	2500	
2	Second	Progeny of single plants		In isolation of 3m
		Self-pollinated	500	
		Cross-pollinated	2500	
		Laboratory evaluation of plant to row progeny		For seed characters
		Self-pollinated	150	
		Cross-pollinated	250	
		Mixing true to type progeny rows		This constitutes nucleus seed
		Self-pollinated	100	
		Cross-pollinated	200	
3	Third	Breeder seed production		Using nucleus seed

Note: Keep some nucleus seed in reserve for use in case of crop failure

2. Sunflower (*Helianthus annuus* L.)

General Information

(i) **Family**: Sunflower belongs to the family *Compositae*. It is an important oilseed crop of global importance. It contains 46 to 52 per cent oil.

(ii) **Global Distribution**: Important sunflower producing countries are USSR, USA, Argentina, China, Australia, France, Spain, Romania, Hungary, Yugoslavia, South Africa and Canada.

(iii) **National Distribution**: In India, it was introduced in 1969 from USSR. Now, it is cultivated in Karnataka, Maharashtra, Tamil Nadu and Andhra Pradesh. It is spreading in Punjab and Haryana states also.

(iv) **Uses**: Sunflower oil is used for human consumption and cake as cattle feed.

Cultivated Species

There are two species of sunflower, *i.e. Helianthus annuus* ($2n = 34$) and *H. tuberosus* which are cultivated as food crops. *H. annuus* is grown as oilseed crop. There are 65 other species, out of which 8 are grown as ornamentals and rest are wild. The cultivated species is of two types, *viz.* oilseed type and non- oilseed type. The non-oilseed types are rich in protein and are grown in USA and some other countries. Non- oilseed types are used whole, dehulled, roasted or raw alone or added to other food.

Origin and Evolution

Archaeological evidences suggest that Central America is the center of origin of sunflower. Nothing is known about the progenitor of sunflower so far.

Reproduction and Pollination

Sunflower is seed propagated and cross pollinated species. Natural cross pollination occurs mainly by insects. Thus sunflower is a highly heterozygous and heterogeneous crop. An isolation distance of 400 m and 200 m is essential for the production of foundation and certified seeds.

Breeding Objectives

Main breeding objectives are higher yield, better quality, wider adaptation, and resistance to biotic (diseases and insects) and abiotic (drought and temperature) stresses.

(i) **Higher Yield**: Major yield components include head size, seeds per head and seed size.

(ii) **Quality Traits**: Quality refers to oil and protein content and their composition.

(iii) **Disease Resistance**: Major diseases include leaf spots, root rot, stem rot, head rot, rust and powdery mildew.

(iv) **Insect Resistance**: *Heliothis*, grasshoppers, jassids and leaf eating cater pillars are major insects.

Breeding Methods

Main breeding methods are mass selection, recurrent selection, heterosis breeding, synthetic breeding, *etc.* Haploids and transgenic plants have been developed through biotechnology.

Breeding Centers

New varieties are developed by State Agricultural Universities and Private seed companies after multi-locational testing in Co-ordinated trials. All India Coordinated

Sunflower Improvement Project is located at the University of Agricultural Sciences, Bengaluru.

Practical Achievements

New varieties are released through Coordinated Project after multi-location testing for 3-5 years. Important sunflower varieties and hybrids released in India are listed below:

1. **Varieties:** EC 68414, EC 68415, Modern, Co 1, Co 2, Surya, SS 56, MSFS 1, MSFS 8 and MSFS 17.
2. **Hybrids:** BSH 1, APSH 11, LDMRSH 1, LDMRSH 3. All these Hybrids are CMS based.

Nucleus Seed Production

Nucleus seed is the base material in seed production chain. The important steps involved in the nucleus seed production of maize are briefly presented as follows:

(i) **Base Material**: Nucleus seed is produced from nucleus seed or breeder seed or sometimes from foundation seed. Select good quality nucleus or breeder or foundation seed for production of nucleus seed. Grow the breeder seed in large plot of at least 0.25 acre. The field of breeder seed production already planted can be used for production of nucleus seed. There is no need of planting separate field for such purpose.

(ii) **Breeding Method Used**: In sunflower plant to row selection is used for production of nucleus seed. Seed of each selected plant is harvested separately and stored in separate envelope.

(iii) **Number of Plants to be selected**: It depends upon the quantity of nucleus seed to be produced. Generally, 500 to 800 individual plants true to type of the variety are selected, selfed and seed of each selected plant is harvested separately.

(iv) **Planting Progeny Rows**: In the second crop season progeny row of each selected plant is planted separately adopting proper isolation distance. Each plant to row progeny is carefully examined from flowering onwards. Off type plants are removed from each progeny row. If variants in a progeny row are many, the entire progeny row is rejected.

(v) **Seed Bulking:** Finally, the seed of 150 to 200 true to the type progeny rows is bulked. This constitutes nucleus seed, which is used for production of breeder seed or nucleus seed in the third crop season. The bulk seed is grown in larger plots to produce breeder seed for further use in seed production chain.

3. Soybean

General Information

(i) **Family**: Soybean belongs to the family *Leguminosea*. It is an important protein cum oil yielding crop of global importance.

(ii) **Global Distribution**: Important soybean producing countries are USA, Brazil, China, Argentina, Canada and Japan. USA alone contributes over 60 per cent to the global production of soybean.

(iii) **National Distribution**: In India, major soybean producing states are M.P., U.P., and Maharashtra.

(iv) **Uses**: In India, both soybean seed and oil are used for human consumption. Soybean contributes 60 per cent to vegetable protein and 30 per cent to edible oil on global basis.

Cultivated Species

Glycine max [2n=40] is the only cultivated species of soybean. This is an annual diploid species. There are 8 wild species of soybean out of which 7 are perennial and one (*Glycine soja*) annual diploid. Wild species are important sources of disease and drought resistance. Soybean is cultivated in tropical and sub-tropical climate.

Origin and Evolution

It is generally accepted that soybean originated in China from where it spread to East and South East Asia, USA and other countries. The wild annual diploid species *Glycine soja* is considered as the progenitor of cultivated soybean. It is also believed that annual form originated from *Glycine tabacina* or *G. tomentella* because these perennial species are also found growing in area of *G. soja*.

Reproduction and Pollination

Soybean is self-pollinated and seed propagated crop. Main points are presented as follows:

(i) **Pollination**: Self-pollination. Flowers are bisexual.

(ii) **Propagation**: Soybean is a seed propagated crop.

(iii) **Isolation Distance**: An isolation distance of 3 meters is safe for the production of foundation and certified seed.

Breeding Objectives

In soybean, main breeding objectives are higher yield, better quality, wider adaptation, and resistance to biotic (diseases and insects) and abiotic (drought and temperature) stresses.

(i) **Higher Yield**: Major yield components are pods per plant, seeds per pod and seed size.

(ii) **Quality Traits**: Quality includes oil and protein content and absence of anti-nutritional factors.

(iii) **Disease Resistance**: Major diseases include pod blight, yellow mosaic, bacterial blight and alternaria leaf spot.

(iv) **Insect Resistance**: *Bihar hairy caterpillar and girdle beetle are the main insects.*

Breeding Procedures

Breeding procedures which are used for soybean improvement are of two types, *viz.* general methods and special methods as follows:

(i) **General Methods**: These methods include introduction, pure line selection, pedigree, back cross and multiline breeding, bulk and single seed descent methods.

(ii) **Special methods**: Such methods include: mutation breeding, distant hybridization and transgenic breeding.

(iii) Somaclonal variation may be helpful in developing resistant genotypes to biotic and abiotic stresses and improving several other traits.

Breeding Centers

Soybean improvement work is carried out by various International and National crop research centers as follows:

(i) **International Centers:** Following Internal crop research centers are involved in soybean improvement.

(a) **AVRDC**: Asian Vegetable Research and Development Center [AVRDC], Taiwan. AVRDC deals with improvement of tomato, Chinese cabbage, sweet potato, mung bean and soybean for Asian region.

(b) **IITA**: International Institute of Tropical Agriculture, Nigeria. IITA deals with improvement of cowpea, soybean, maize, rice, root tubers and plantation crops.

(ii) **National Centers:** In India, soybean improvement work is carried out by following organizations:

(a) Directorate of Soybean Research, Indore, Madhya Pradesh. The new varieties are released through coordinated project after multi-location testing for 3-5 years.

(b) State Agricultural Universities located in soybean growing regions.

Practical Achievements

In India, several improved varieties of soybean have been developed for different states. Eight varieties (Bragg, Clark 63, Davis, Hardee, Improved Pelicon, Jupiter, Lee and Monetta) have been released through direct introduction. Some high yielding varieties have been developed by Pantnagar, Jabalpur, Ludhiana and other Agricultural universities and also by IARI, New Delhi. These varieties include PK 262, PK 308, PK 327, PK 416, PK 472, Shilajeet and Pant Soybean 564 (Pantnagar); Durga, Gaurav, JS 75-46, JS 80-21, (*Jabalpur); SL 4 and SL 96 (Ludhiana); Pusa 16 Pusa 20 and Pusa 24 (IARI, New Delhi); VL Soya 2 (Almora), Shivalik (Palampur); and MACS 13 and MACS 58 (Poona).

Nucleus Seed Production

Nucleus seed is the base material in seed production chain. The important steps involved in the nucleus seed production of soybean are briefly presented as follows:

(i) **Base Material**: Nucleus seed is produced from nucleus seed or breeder seed or sometimes from foundation seed. Select good quality nucleus or breeder or foundation seed for production of nucleus seed. Grow the breeder seed in large plot of at least 0.25 acre. The field of breeder seed production already planted can be used for production of nucleus seed. There is no need of planting separate field for such purpose.

(ii) **Breeding Method Used**: In soybean, plant to row selection is used for production of nucleus seed. Seed of each selected plant is harvested separately and stored in separate envelope.

(iii) **Number of Plants to be selected**: It depends on the quantity of nucleus seed to be produced. Generally, 400 to 500 single plants true to type of the variety are selected and seed of each selected plant is harvested separately.

(iv) **Planting Progeny Rows**: In the second crop season progeny row of each selected plant is planted separately adopting proper isolation distance. Each plant progeny is carefully examined from flowering onwards. Off type plants are removed from each progeny row. If variants in a progeny row are many, the entire progeny row is rejected.

(v) **Bulking Seed:** The seed of all true to the type progeny rows is bulked. This constitutes nucleus seed, which is used for production of breeder seed or nucleus seed in the third crop season. The bulk seed is grown in larger plots to produce breeder seed for further use in seed production chain.

TABLE 14.1: Steps in Nucleus Seed Production of Soybean

Sl.No.	*Season*	*Particulars*	*Details*	*Remarks*
1	First	Base Population to be raised	0.25 Acre	In disease free field
		Single plants to be selected	500	Tag selected ears
2	Second	Plant to rows progeny	500	In isolation of 5m
		Laboratory evaluation of progeny	250	For seed characters
		Bulking equal number of seed from best progenies.	200	This constitutes nucleus seed
3	Third	Breeder seed production		Using nucleus seed

Note: Keep some nucleus seed in reserve for use in case of crop failure.

4. Groundnut

General Information

(i) **Family:** Ground nut belongs to the family *Leguminosea*. It is an important oil yielding crop of global importance.

(ii) Global Distribution: Peanut is a tropical crop, but its cultivation extends to both tropical and sub- tropical regions of the world. Principal peanut producing countries are India, China, USA, Africa [Senegal and Nigeria] and South and South East Asia [Myanmar, Malayasia Pakistan, and Sri Lanka].

(iii) National Distribution: In India, major peanut producing states are Gujarat, Andhra Pradesh, Karnataks, Tamil Nadu, Maharashtra, Madhya Pradesh, Rajasthan, Uttar Pradesh, Odisha and Punjab.

(iv) Uses: Kernels are rich in oil, protein, vitamins and minerals and are used raw or roasted for human consumption. Leaves and cake are used as cattle feed shells as fuel.

Cultivated Species

Arachis hypogaea [2n=40] is the only cultivated species of peanut. This is an annual tetraploid species. This species has two sub-species, *viz. Arachis hypogaea ssp hypogea* and *A. hypogaea ssp fastigiata*. The former sub-species has no floral axis on the main stem, while the latter has. Each sub-species has two types as follows:

(i) Sub-species *hypogaea*: It has two types, *viz. hypogaea* and hirsute. The variety hypogaea has lesser hair and shorter branches than hirsute.The variety *hypogaea* is of two types, *viz.* small poded with spreading (runner) growth habit called runner, and large poded with erecthabit called Virginia. The hirsute is not much cultivated.

(ii) Sub-species *fastigiata*: It is of two types, *viz. fastigiata* or *valencia* and *vulgaris* or Spanish. The Valencia is large podded and Spanish has small pods.

There are about 15 wild species of peanut which are used in interspecific gene transfer for resistance to biotic and abiotic stresses.

Origin and Evolution

It is generally accepted that peanut originated in Brazil from where it spread to other countries. The tetraploid wild species *Arachis monticola* [2n= 4x=40] is considered to be the progenitor of cultivated peanut. The wild species *Arachis monticola* has several similarities with cultivated species. Moreover, it can be easily crossed with cultivated peanut.

Reproduction and Pollination

Cultivated peanut is self-pollinated and seed propagated crop. Main points are presented as follows:

(i) Pollination: Peanut is a self-pollinated species. The percentage of natural cross pollination is practically negligible.

(ii) Propagation: Peanut is a seed propagated crop. Some wild species propagate vegetatively usually by rhizomes.

(iii) Isolation Distance: An isolation distance of 3 meters is safe for the production of foundation and certified seed.

Breeding Objectives

In peanut, main breeding objectives are higher yield, better quality, wider adaptation, and resistance to biotic (diseases and insects) and abiotic (drought and temperature) stresses.

(i) **Higher grain Yield:** Major yield components are pods per plant, kernels per pod, kernel size and shelling per cent.

(ii) **Quality Traits:** Quality includes seed oil and protein content and composition, flavor and milling quality.

(iii) **Disease Resistance**: Major diseases include leaf spot, dry root rot, Fusarium wilt, mosaic virus and rosette.

(iv) **Insect Resistance:** Important insects are red caterpillar, Bihar hairy caterpillar, leaf miner, pod borer and pod sucking bugs.

Breeding Procedures

Breeding procedures which are used for peanut improvement are of three types, *viz.* general methods, special methods and population improvement procedures as follows:

(i) **General Methods**: These methods include introduction, pure line selection, pedigree, back cross and multiline breeding, bulk and single seed descent methods.

(ii) **Special methods**: Such methods include: mutation breeding, distant hybridization and transgenic breeding.

(iii) **Population Improvement**: Three breeding procedure, *viz.* biparental mating, disruptive mating and diallel selective mating are useful in breaking linkage blocks and creating vast genetic variability in a population.

(iv) **Biotechnology**: Biotechnological tools such as embryo rescue and tissue culture has made it possible to utilize genes from wild species.

Breeding Centers

Peanut improvement work is carried out by various International and National crop research centers as follows:

(i) **International Centers:** International Crop Research Institute for Semi-arid Tropics [ICRISAT], Hyderabad is involved in peanut improvement.

(ii) **National Centers:** In India, soybean improvement work is carried out by following organizations

(a) Directorate of Groundnut Research, Junagarh, Gujarat.

(b) State Agricultural Universities located in peanut growing regions.

The new varieties are released through coordinated project after multi-location testing for 3-5 years.

Practical Achievements

In India, more than 60 improved varieties of peanut have been released for cultivation in different. states. List of important peanut varieties released in Spanish, Valencia and Virginia groups in India is presented as follows:

(i) **Spanish Type**: MH 1, SG 84,GG 2, J 11, Jyoti, TG 17, Javan, Kisan, ICGS 11, Co 1, Co 2, TMV 5, TMV 7, TMV 9, TMV 12, *etc.*

(ii) **Valencia Type**: MH 2, Kopergaon 3, TMV 3, *etc.*

(iii) **Virginia Type**: T 28, T 64, M 145, M197, TMV 6, TMV 8, TMV 10, Kausal, Kadri 2 and Kadri 3 are semi-spreading types. Spreading types include M 13, M37, M335, GG 11, TMV 1, TMV 3, TMV 4, S 230, Kadri 71-1, *etc.*

Nucleus Seed Production

Nucleus seed is the base material in seed production chain. The important steps involved in the nucleus seed production of groundnut are briefly presented as follows:

(i) **Base Material**: Nucleus seed is produced from nucleus seed or breeder seed or sometimes from foundation seed. Select good quality nucleus or breeder or foundation seed for production of nucleus seed. Grow the breeder seed in large plot of at least 0.25 acre. The field of breeder seed production already planted can be used for production of nucleus seed. There is no need of planting separate field for such purpose.

(ii) **Breeding Method Used**: In groundnut, plant to progeny selection is used for production of nucleus seed. Seed of each selected single plant is harvested separately and stored in separate envelope.

(iii) **Number of Plants to be selected**: It depends on the quantity of nucleus seed to be produced. Generally, 100 to 150 single plants true to type of the variety are selected and seed of each selected plant is harvested separately.

(iv) **Planting Progeny Rows**: In the second crop season progeny row of each selected single plant is planted separately adopting proper isolation distance. Each plant to row progeny is carefully examined from flowering onwards. Off type plants are removed from each progeny row. If variants in a progeny row are many, the entire progeny row is rejected.

(v) **Laboratory Evaluation:** In the laboratory, seed of all progeny of single plants are evaluated in terms of shape, size, colour and lustre.

(vi) **Bulking Seed:** The seed of all true to the type progeny rows is bulked. This constitutes nucleus seed, which is used for production of breeder seed or nucleus seed in the third crop season. The bulk seed is grown in larger plots to produce breeder seed for further use in seed production chain.

5. Castor

Castor is an important oilseed crop. It is monoecious having pistillate flowers [female] on the upper portion of the race and staninate flowers on the lower part. Castor is highly cross-pollinated crop. For production of breeder and nucleus seed, an isolation distance of 1500m for varieties and male parents and 2000m female parent is required.

Nucleus Seed Production

Nucleus seed is the base material in seed production chain. The important steps involved in the nucleus seed production of castor are briefly presented as follows:

(i) **Base Material**: Nucleus seed is produced from nucleus seed or breeder seed or sometimes from foundation seed. Select good quality nucleus or breeder or foundation seed for production of nucleus seed. Grow the breeder seed in large plot of at least 0.25 acre. The field of breeder seed production already planted can be used for production of nucleus seed. There is no need of planting separate field for such purpose.

(ii) **Breeding Method Used**: In castor, plant to progeny selection is used for production of nucleus seed. Seed of each selected single plant is harvested separately and stored in separate envelope.

(iii) **Number of Plants to be selected**: It depends on the quantity of nucleus seed to be produced. Generally, 200 to 500 single plants true to type of the variety are selected and seed of each selected plant is harvested separately.

(iv) **Planting Progeny Rows**: In the second crop season progeny row of each selected single plant is planted separately adopting proper isolation distance. Each plant to row progeny is carefully examined from flowering onwards. Off type plants are removed from each progeny row. If variants in a progeny row are many, the entire progeny row is rejected.

(v) **Laboratory Evaluation:** In the laboratory, seed of all progeny of single plants are evaluated in terms of shape, size, colour and lustre.

(vi) **Seed Bulking:** The seed of all true to the type progeny rows is bulked. This constitutes nucleus seed, which is used for production of breeder seed or nucleus seed in the third crop season. The bulk seed is grown in larger plots to produce breeder seed for further use in seed production chain.

Summary

Important oil yielding crops include rapeseed and mustard, ground nut, soybean, sunflower, *etc.* An area of plant breeding which deals with principles and methods of breeder seed production is referred to as maintenance breeding or varietal maintenance technology. Main points related to base material, breeding procedure used, main focus, plant part used, progenies grown, isolation distance, selection of field, crop inspection and evaluation of selected progenies for maintenance breeding [nucleus seed production] of rapeseed and mustard, ground nut, soybean, sunflower, *etc.* have been discussed.

QUESTIONS

1. Describe briefly genetic origin, species composition and breeding objectives of Rapeseed and mustard.
2. Discuss in brief breeding procedure used for genetic improvement of Rapeseed and mustard.
3. Explain in short practical achievements of Rapeseed and mustard breeding in India.
4. Describe in brief important Rapeseed and mustard breeding centres and their role in varietal improvement.
5. Describe briefly genetic origin, species composition and breeding objectives of Sunflower.
6. Discuss in brief breeding procedure used for genetic improvement of Sunflower crop.
7. Explain in short practical achievements of Sunflower breeding in India.
8. Describe briefly genetic origin, species composition and breeding objectives of Soybean in India.
9. Discuss in brief breeding procedure used for genetic improvement of Soybean crop.
10. Explain in short practical achievements of Soybean breeding in India.
11. Describe briefly genetic origin, species composition and breeding objectives of Peanut in India.
12. Discuss in brief breeding procedure used for genetic improvement of Peanut crop.
13. Explain in short practical achievements of Peanut breeding in India.
14. Discuss in brief role of International and National Crop Research Centres in crop breeding
15. Describe briefly the procedure of nucleus seed production in mustard.
16. Explain briefly the procedure of nucleus seed production in Soybean.
17. Discuss briefly the procedure of nucleus seed production in Sunflower.
18. Describe briefly the procedure of nucleus seed production in groundnut.
19. Describe briefly the procedure of nucleus seed production in castor.

* * * *

Chapter 15

Nucleus Seed Production of Pulse Crops

Introduction

Those Leguminous species whose seeds are used for human consumption are referred to as pulses or grain legumes (Roberts, 1970). Important pulse crops include chickpea, field pea, pigeon pea, black gram, green gram, lentil, horse gram, moth bean, field bean, *khesary*, cowpea, cluster bean, *etc.* Pulse crops belong to the family *Fabaceae* (old *Leguminosae*).

Types

Pulse crops are classified in various ways as presented in the Table 15-1.

TABLE 15.1: Classification of Pulse Crops

Sl.No.	*Basis of Classification*	*Categories*	*Examples of Pulse Crops*
1	Utilization	Major or Primary Species	Chickpea, pea, lentil, pigeon pea, green gram, black gram and cowpea.
		Minor or Secondary Species	Horse gram, moth bean, field bean, *khesary*, and cluster bean.
2	Temperature tolerance	Cool season Species	Chickpea, pea and lentil.
		Warm season Species	Pigeon pea, green gram, black gram and cowpea.
3	Growing season	Rabi season	Chickpea, pea and lentil.
		Kharif season	Pigeon pea, green gram, black gram and cowpea.

(i) Cool Season Pulses: Such species include chickpea, field pea, lentil and *khesari. These* are cool season pulses of subtropical and mild temperate regions.

(ii) Warm season Pulses: Such species include Pigeon pea, green gram, black gram and cowpea. Based on utilization, pulse crops are again of two types as follows:

(a) Primary Pulses: This group includes chickpea, pea, lentil (cool season), pigeon pea, green gram, black gram and cowpea (warm season). These are also known as major pulse crops.

(b) Secondary Pulses: This group includes horse gram, moth bean, field bean, *khesary*, and cluster bean. These are also known as minor pulses.

A brief description and breeding of primary pulses is presented here. Major producing countries and Indian states for primary pulses are given in Table 15.2.

TABLE 15.2: Distribution of Primary Pulse Crops

Sl.No.	*Pulse Crop*	*Cultivated Species*	*Major growing Countries*	*Major growing Indian States*
1	Chick pea	*Cicer arietinum*	India, Pakistan, Turkey, Ethiopia and Burma	M.P., Rajasthan, U.P. and Haryana
2	Field pea	*Pisum sativum*	China, India, Pakistan, USA, and Canada,	U.P., M.P. Rajasthan and Bihar
3	Lentil	*Lens culinaris*	India, Pakistan, Egypt and Greece	U.P., M.P. Bihar and West Bengal.
4	Pigeon pea	*Cajanus cajan*	India, Uganda, Kenya, West Indies, Burma, *etc.*	Maharashtra, U.P., M.P. Karnataka, Gujarat and Bihar.
5	Black gram	*Vigna mungo*	India, Pakistan, Bangladesh, Sri Lanka, Philippines, Taiwan, Thailand, Nepal, *etc.*	M.P. Maharashtra, Tamil Nadu, U.P., A.P., *etc.*
6	Green gram	*Vigna radiata*	India, Pakistan, Bangladesh, Sri Lanka, Philippines, Taiwan, Thailand, Nepal *etc.*	Odisha, A.P., M.P., Maharashtra and Rajasthan.
7	Cow pea	*Vigna anguiculata*	Nigeria, Niger, Ghana, Kenya, Uganda, Tanzania, India, Sri Lanka, Burma, *etc.*	Rajasthan, M.P., U.P., Maharashtra and Gujarat.

Above pulses contain 18 to 27 per cent protein. The lowest protein is found in chickpea and field pea (18-20 per cent) and the highest in green gram and cowpea (24-27 per cent). Dry seeds are used as Dal, green pods as vegetables and leaves and stalks as cattle feed. Moreover, pulses fix 30 to 50 kg nitrogen per hectare in the soil per season.

Origin and Cultivated Species

The cultivated species, center of origin and probable progenitors of primary pulse crops are given in Table 15.3.

TABLE 15.3: Place of Origin and Probable Progenitor of Primary Pulse Crops

Sl.No.	*Name of Crop*	*Cultivated Species*	*Place of Origin*	*Progenitor*
1	Chick pea	Cicer arietinum	South Turkey and Syria	*Cicer reticulatum*
2	Field pea	Pisum sativum	Mediterranea, Ethiopia and Near East	Not known
3	Lentil	Lens culinaris	Near East	*Lens orientales*
4	Pigeon pea	*Cajanus cajan*	Africa and India	*Cajanus cajanifolius*
5	Black gram	*Vigna mungo*	Indian sub-continent	*Vigna radiata* var *sublobatis*
6	Green gram	*Vigna radiata*	India	*Vigna radiata* var *sublobata*
7	Cow pea	*Vigna anguiculata*	Africa (PC), India, China (SC)	Not known

Reproduction and Pollination

All primary pulses are seed propagated and self-pollinated. Other points are as follows:

(i) Flowers: Flowers are bisexual and amenable for crossing.

(ii) Pollination: All pulses are self-pollinated. Pigeon pea is often cross pollinated species in which natural cross pollination up to 7 per cent has been reported under Indian conditions.

(iii) Propagation: All pulses are seed propagated.

(iv) Isolation Distance: Isolation distance of 20 m. in chickpea, field pea, lentil, black gram and green gram; 50 m. in cowpea; and 200 m. in pigeon pea is required for the production of foundation and certified seeds.

Breeding Objectives

In grain legumes, major breeding objectives are, high yield, wider adaptation, shattering resistance, better quality, early maturity, resistance to biotic and abiotic stresses as follows:

(i) Higher grain yield: Major yield components are, number of pods per plant, pod length, seeds per pod and seed size.

(ii) Better quality: Main quality characters are seed color, seed size, protein content, methionine content, and low amount of anti-nutritional substance.

(iii) Disease Resistance: *Fusarium* wilt and rust are major diseases of chickpea, pea, and lentil. Yellow mosaic, bacterial blight, leaf spot, powdery mildew and root rot are common in pigeon pea, black gram, green gram and cowpea. In Pigeon pea, *Fusarium* wilt and sterility mosaic are also major diseases.

(iv) **Insect Resistance**: Pod borer, cutworm and aphids in Rabi pulses; and pod borer, hairy cater pillar and jassids in *Kharif* pulses (except pigeon pea) are the major insect pests.

Breeding Procedures

Breeding procedures which are used for improvement of pulse crops are of three types, *viz.* general methods, special methods and population improvement procedures as follows:

(i) **General Methods**: These methods include introduction, pure line selection, pedigree, back cross and multiline breeding, bulk and single seed descent methods.

(ii) **Special methods:** Such methods include: mutation breeding, distant hybridization, heterosis breeding and transgenic breeding. Singh (1991) has reviewed the mutation breeding work on pulse crops. Mutation breeding has helped in developing improved varieties in various pulse crops such as chickpea (Pusa 408, Pusa 413, Pusa 417 and Kiran), Lentil (Arun), field pea (Hans), Pigeonpea (Co 5, Co 3, TAT 5 and Trombay), blackgram (Co 4), greengram (Co 4, ML 26-1-3, Pant Mung 2 and TAP 7) and cowpea (V 16, V 3, V 38 and V 240). In Pigeon pea, heterosis breeding is also used. Biotechnology will supplement to conventional methods in pulse crops.

(iii) **Population Improvement Procedures**: Such procedures include recurrent selection and diallel selective mating systems. These are used for specific purposes.

Breeding Centers

There are two types of pulse breeding centers as follows:

(i) **International Centers:** Following International Research Centers are engaged in improvement of pulse crops:

(a) **ICRISAT**: International Crop Research Institute for Semiarid Tropics, Hyderabad, India is actively engaged in improvement of pigeon pea and chickpea (both desi and Kabuli).

(b) **ICARDA:** International Centre for Agricultural Research in Dryland Areas (ICARDA) works on improvement of lentil and Kabuli chickpea.

(c) **AVRDC:** Asian Vegetable Research and Development Centre, Taiwan deals with breeding of mung bean and maintains global gene pool of this pulse.

(d) **CIAT**: International Centre for Tropical Agriculture (CIAT), Colombia deals with breeding of French bean.

(e) **IITA**: International Institute of Tropical Agriculture (IITA), Nigeria deals with improvement of Cowpea.

(ii) National Centers: In India, pulse breeding work is carried out by following organizations:

(a) Indian Institute of Pulses Research Kalyanpur, Kanpur The new varieties are released through coordinated project after multi-location testing for 3-5 years.

(b) State Agricultural Universities located in pulses growing regions.

Improvement of grain type cowpea has been included in the ICAR Coordinated Project on underutilized unexploited crops. Vegetable types are dealt with by Indian Institute of Vegetable Research, Varanasi and work on fodder legumes is carried out at Indian Grassland and Fodder Research Institute, Jhansi.

Practical Achievements

In India, several high yielding varieties of grain legumes have been released for different states. ICRISAT has developed the first pigeon pea hybrid (ICPH 8) in the world for commercial cultivation in India. This hybrid has been developed using genetic male sterility. It is early maturing (140 days) and performs well under drought and high moisture conditions. This hybrid has been recommended for Maharashtra and Gujarat states. A list of important varieties released in various pulse crops in India is given below:

(i) **Pigeon pea:** T 21, T 7, T 17, Prabhat, UPAS 120, Manak, Vishaka, ICPL 87, ICPL 151, LRC 30, LRC 36, Pusa 84, Pusa 85, *etc.*

(ii) **Chick pea:** T 3, T 4, T 5, Pant 114, K 850, H 208, H 355, Pusa 209, Pusa 240, Pusa 244, Pusa 256, Pusa 413, Pusa 417, Pusa 212, Avrodhi, *etc.*

(iii) **Lentil:** T 8, T 36, Pant 406, Pant L 639, L 9-12, Lens 4076, Malika, Pusa 4, Pusa 6, *etc.*

(iv) **Pea:** Type 163, Pant P 5, Rachna, DMR 11, Pant P 5, Hans, VL 1, B 12, *etc.*

(v) **Black gram:** T 9, T 27, T 65, PS 1, Pant 419, 430, Khargone 3, Co 1, Co 2, Co 3, Co 4, Co 5, *etc.*

(vi) **Green gram:** P 37, P 310, PS 16, Pusa 105, Pusa Baisakhi, T 44, T 51, K 851, Plant Moong 2, Plant Moong 3, *etc.*

(vii) **Cowpea:** Pusa 152, Pusa Sawani, V 16, V 240, Co 1, Co 2, Co 3, Co 4, K 11, K 14, T 2, JC 5, JC 10, *etc.*

The yield of pulse crops is lower than cereals. Main reasons of low yield of pulse crops are: (i) long continued natural selection, (ii) cultivation on poor lands, (iii) inadequate application of fertilizers, (iv) inadequate plant protection measures, and (v) rain-fed cultivation.

Nucleus Seed Production

Nucleus seed is the base material in seed production chain. The important steps involved in the nucleus seed production of maize are briefly presented as follows:

(i) **Base Material**: Nucleus seed is produced from nucleus seed or breeder seed or sometimes from foundation seed. Select good quality nucleus or breeder or foundation seed for production of nucleus seed. Grow the breeder seed in large plot of at least 0.25 acre. The field of breeder seed production already planted can be used for production of nucleus seed. There is no need of planting separate field for such purpose.

(ii) **Breeding Method Used**: In pulse crops single plant selection is used for production of nucleus seed. Single plants true to the type are selected after flower initiation. Seed of each selected single plant is harvested separately and stored in separate envelope.

(iii) **Number of Plants to be selected**: It depends on the quantity of nucleus seed to be produced. Generally, 100 to 150 single plants true to type of the variety are selected and seed of each selected single plant is harvested separately.

(iv) **Planting Progeny Rows**: In the second crop season progeny row of each selected plant is planted separately adopting proper isolation distance. Each plant to row progeny is carefully examined from flower initiation onwards. Off type plants are removed from each progeny row. If variants in a progeny row are many, the entire progeny row is rejected.

(v) **Laboratory Evaluation:** In the laboratory, seed of all progeny of single plants are evaluated in terms of shape, size, colour and lustre.

TABLE 15.4: Steps in Nucleus Seed Production of Chick Pea

Sl.No.	*Season*	*Particulars*	*Details*	*Remarks*
1	First	Base Population to be raised	0.25 Acre	In disease free field
		Single plants to be selected	500	Tag selected ears
2	Second	Plant to progeny rows	500	In isolation of 5m
		Laboratory evaluation of progeny	250	For seed characters
		Bulking equal number of seed from best progenies.	200	This constitutes nucleus seed
3	Third	Breeder seed production		Using nucleus seed

Note: Keep some nucleus seed in reserve for use in case of crop failure.

TABLE 15.5: Steps in Nucleus Seed Production of Pigeon Pea

Sl.No.	*Season*	*Particulars*	*Details*	*Remarks*
1	First	Base Population to be raised	0.25 Acre	In disease free field
		Single plants to be selected	500	Tag selected ears
2	Second	Plant to progeny rows	500	In isolation of 200m
		Laboratory evaluation of progeny	250	For seed characters
		Bulking equal number of seed from best progenies.	200	This constitutes nucleus seed
3	Third	Breeder seed production		Using nucleus seed

Note: Keep some nucleus seed in reserve for use in case of crop failure.

(vi) **Seed Bulking:** The seed of all true to the type progeny rows is bulked. This constitutes nucleus seed, which is used for production of breeder seed or nucleus seed in the third crop season. The bulk seed is grown in larger plots to produce breeder seed for further use in seed production chain.

TABLE 15.6: Steps in Nucleus Seed Production of Field Pea

Sl.No.	*Season*	*Particulars*	*Details*	*Remarks*
1	First	Base Population to be raised	0.25 Acre	In disease free field
		Single plants to be selected	500-1000	Tag selected ears
2	Second	Plant to rows progeny	500	In isolation of 5m
		Laboratory evaluation of progeny	250	For seed characters
		Bulking equal number of seed from best progenies.	200	This constitutes nucleus seed
3	Third	Breeder seed production		Using nucleus seed

Note: Keep some nucleus seed in reserve for use in case of crop failure.

Summary

Important pulse crops include chickpea, field pea, pigeon pea, black gram, green gram, lentil, horse gram, moth bean, field bean, *khesary*, cowpea, cluster bean, *etc.* Pulse crops belong to the family *Fabaceae* (old *Leguminosae*). An area of plant breeding which deals with principles and methods of breeder seed production is referred to as maintenance breeding or varietal maintenance technology. Main points related to base material, breeding procedure used, main focus, plant part used, progenies grown, isolation distance, selection of field, crop inspection and evaluation of selected progenies for maintenance breeding [nucleus seed production] of chickpea, field pea, pigeon pea, black gram, green gram, lentil, horse gram, moth bean, field bean, *khesary*, cowpea, cluster bean, *etc.* are discussed.

QUESTIONS

1. **Describe briefly genetic origin, species and breeding objectives of primary pulse crops**
2. **Discuss in brief breeding procedure used for genetic improvement of primary pulse crops**
3. **Explain in short practical achievements of pulse crop breeding in India.**
4. **Describe in brief important pulse crop breeding centres and their role in varietal improvement.**
5. **Describe briefly probable progenitors of primary pulse crops.**

6. **Write Short notes on the following:**
   (i) Primary Pulse crops
   (ii) Secondary Pulse Crops
   (iii) Cool season pulse crops
   (iv) Warm season pulse crops
7. **Describe briefly the procedure of nucleus seed production in cool season pulse crops.**
8. **Explain briefly the procedure of nucleus seed production in warm season pulse crops.**
9. **Discuss briefly the procedure of nucleus seed production in Pigeon pea.**

* * * *

Chapter 16

Nucleus Seed Production of Fiber Crops

Introduction

Important commercial Fibre crops include cotton, jute, ramie, *etc.* A brief account of distribution, cultivated species, origin and evolution, reproduction and pollination, breeding objectives, breeding procedures, breeding centers, practical achievements and nucleus seed production of these crops is presented as follows:

1. Cotton (*Gossypium* sp.)

General Information

(i) **Family:** Cotton belongs to the family *Malvaceae* and produce spinnable seed coat fibers. There are two types of fibres, *viz.* long fibres called lint and small fibres called linters or fuzz.

(ii) **Global Distribution**: Cotton is grown in tropical and subtropical regions of more than 80 countries of the world. The major cotton growing countries are China, USA, India, Pakistan, Uzbekistan, Egypt, Sudan, Turkey, *etc.*

(iii) **National Distribution**: In India, there are nine major cotton growing states, *viz.* Punjab, Haryana, Rajasthan (North Zone), Maharashtra, Madhya Pradesh, Gujarat (Central Zone), Andhra Pradesh, Karnataka and Tamil Nadu (South Zone).

(iv) **Types**: Cotton is of two typed as Old World cottons and New World cotton. Cotton is also classified as diploid cotton and tetraploid cotton.

(v) **Uses**: Cotton is a fibre, oil and protein yielding crop. The lint is used in the textile mills, linters for various industrial uses, refined cotton seed oil is used for human consumption and the cake as manure and livestock feed.

Cultivated Species

In the genus ***Gossypiun*** about 50 species have been identified so far. Out of these four species, *viz., Gossypium hirsutum, G. barbadense, G. arboreum* and *G. herbaceum* are cultivated and rest are wild or uncultivated. The first two species are tetraploid [2n =52] and last two species are diploid [2n =26]. Tetraploid cottons are also known as New World cottons and diploid cottons are also known as Old World cottons or Asiatic cottons. India is the only country where all the four cultivated species are commercially planted. The *G. hirsutum* is also known as upland cotton or American cotton. *G. barbadense* is also known as Egyptian cotton or Peruvian cotton or Tanguish cotton or Sea Island cotton.

Origin and Evolution

Central Africa is considered as the centre of origin of the genus *Gossypium,* where *G. africanum* is still found growing in the wild state. Genetic origin of four cultivated species is discussed as follows:

(i) **Diploid Cottons:** It is believed that diploid species of cotton originated first and tetraplois species later on from a cross between two diploid species. The wild linted species *Gossypium africanum* is considered as the progenitor of all cultivated species of cotton. *Gossypium africanum* reached undivided India from South Africa through travelers, traders and explorers and after differentiation gave birth to two diploid cultivated species, *viz. Gossypium arboretum* and *Gossypium herbaceum.*

(ii) **Tetraploid Cottons:** There are two species of tetraploid cultivated cotton, *viz. Gossypium hirsutum* and *G. barbadense*. The genome constitution of both these species is AD. It is believed that A genome of tetraploid cotton has come from Asian cultivated species and D genome from American lintless wild diploid species (Skovsted, 1934). Skovsted (1934) first pointed out that the New World tetraploid consisted of A and D sub-genomes. Later on origin of tetraploid cotton was explanined by other workers. There are two views about the genetic origin of tetraploid cotton as discussed below.

(1) **Beasley and Harland View**: In 1940, Beasley and Harland independently observed that A genome of upland cotton has come from Asian cultivated diploid species *Gossypium arboretum* and D genome from American wild lintless diploid species *G. thurberi.* The cross between these two species took place in nature followed by chromosomal doubling. This is represented as shown in Figure 16.1.

(2) **Phillips (1963) View:** He reported that A genome of upland cotton has come from African linted wild diploid species *Gossypium africanum* and D genome from American wild lintless diploid species *G. raymondii.* The cross between these two species took place in nature followed by chromosomal doubling. This is represented as shown in Figure 16.2.

Parents	*Gossypium arboretum*	x *Gossypium thurberi*
Genome	AA [2n=26, large]	DD [2n =26, small
F_1	AD [Sterile]	
Chromosome doubling	AA DD Fertile = Like *G. hirsutum*	
	[4n= 52, 26 large and 26 small]	

FIGURE 16.1: Probable Origin of *G. hirsutum* as Proposed Independently by Beasley (1940) and Harland (1940).

Parents	*Gossypium africanum*	*x Gossypium raymondii*
Genome	AA [2n=26, large]	DD [2n =26, small
F_1	AD [Sterile]	
Chromosome doubling	AA DD Fertile =Like *G. hirsutum*	
	[4n= 52, 26 large and 26 small]	

FIGURE 16.2: Probable Origin of *G. hirsutum* as Proposed Phillips (1963).

Cytological, biochemical [electrophoretic studies] and molecular investigations have clearly indicated that A genome of *G. africanum* is closer to A genome of tetraploid cotton than that of *G. arboretum*. Similarly, D genome of *G. raymondii* is closer to D genome of tetraploid cotton than that of *G. thurberi*. Thus *G. africanum* and *G. raymondii* are progenitors of tetraploid cotton. The second view is widely accepted.

Hutchinson, Silow and Stephens (1947) presented classification of the genus Gossypiun on the basis of cytological, genetic, geographical and archeological evidences. They divided all the cultivated and wild species of cotton in to eight sections, *viz.*, Sturtiana, Erioxyla, Klotzschiana, Thurberana, Anomala, Stockiana, Herbacea and Hirsuta. Later on Phillips added one more section *i.e.* Longicalyciana.

Reproduction and Pollination

Cotton is a self-pollinated and seed propagated crop. Main points are presented as follows:

(i) **Flowers**: Flowers are bisexual. Flowers are showy and attracts insects.

(ii) **Pollination**: Cotton is basically a self-pollinated crop. Cross pollination varies from nil to 60 per cent under different climatic conditions.

(iii) **Apomixis**: In cotton semi-gamy is reported

(iv) **Propagation**: Sexually produced seeds are used for propagation.

(v) **Branching**: Cotton crop has two types of branches, *viz.* monopodia and sympodia.

(vi) Isolation Distance: An isolation distance of 50 m and 30 m is required for the foundation of and certified seed.

Breeding Objectives

In cotton, major breeding objectives are higher yield, better fibre quality, early maturity, wider adaptability and resistance to biotic and abiotic stresses as follows:

(i) Higher Lint Yield: Major yield components are: number of bolls per plant and boll weight.

(ii) Fibre Quality: Fibre qualities include fibre length, fibre strength, fibre fineness, fibre maturity, *etc.*

(iii) Disease Resistance: Major diseases include *Fusarium* wilt, *Verticillium* wilt and bacterial blight.

(iv) Insect Resistance: Important insects include boll worms, jassids, aphids, white fly, thrips, mites, *etc.*

Breeding Procedures

Breeding procedures which are used for cotton improvement are of three types, *viz.* general methods, special methods and population improvement procedures as follows:

(i) General Methods: These methods include introduction, pure line selection, pedigree, back cross and multiline breeding, bulk and single seed descent methods.

(ii) Special methods: Such methods include: mutation breeding, distant hybridization, heterosis breeding and transgenic breeding.

(iii) Population Improvement Procedures: Such procedures include recurrent selection and diallel selective mating systems. These are used for specific purposes.

Biotechnology has helped in developing transgenic cotton with resistance to *Helicoverpa.* The resistant gene has been transferred from bacteria *Bascillus thuringiensis* into cotton plant by Mansanto Seed Company in USA.

Breeding Centers

There are two types of cotton breeding centers as follows:

(i) International Centers: No international research center for cotton breeding has been established so far. There is International Cotton Advisory Committee (ICAC) which advises on various matters of cotton research and marketing.

(ii) National Centers: In India, cotton breeding work is carried out by following organizations:

(a) Central Institutes for Cotton Research, Nagpur. The new varieties are released through coordinated project after multi-location testing for 3-5 years.

(b) State Agricultural Universities located in cotton growing regions.

(c) Some private seed companies.

Practical Achievements

There are three main achievements in the improvement of cotton crop in India after independence. These are: (1) acclimatization of G. *barbadense* cotton, (2) improvement of *hirsutum* and (3) development of hybrids. Egyptian cotton got acclimatized in India in 1955. The yield of *hirsutum* varieties has been significantly improved. India is the pioneer country in the world for the commercial cultivation of hybrid cotton. Several cotton hybrids have been developed in India. However, most of them have been evolved by conventional method *i.e.* hand emasculation and pollination. One hybrid *viz.* Suguna has been developed through the use of genetic male sterility and two (PKVHY 3 and PKVHY 4) through the use of cytoplasmic genic male sterility so far. *G. hirsutum* and *G. arboreum* are cultivated in all cotton growing states. However, *G. barbadense* is cultivated in few pockets of Tamil Nadu and Andhra Pradesh, and *G. herbaceum* in some parts of Gujarat and Karnataka. The important varieties and hybrids of cotton are listed below:

1. Varieties

G. hirsutum. LH 900, LH 1556, F 846, F 1378, F 1861, HS 6, H 874, H 1098, HS 182, H 1117 RST 9, RS 875, 810 (North zone), Khandwa 3, Vikram, PKV 081, Rajat, LRA 5166, LRK 516 (Central zone), Abadita, Sharada, Sahana, MCU 12, MCU 5 VT, MCU 10, and LRA 5166. (South zone)

G. arboreum. LD 491, LD 694, RG 18, HD 107, HD 123, LD 327, RG 8 (North zone), AKH 4, AKA 8401, Maljari, AKA 5, AKA 7, PA 255, PA 402 (Central zone) K 9, K 10, K 11 (South zone).

2. Hybrids

(*a*) **Interspecific hybrids.** Varalaxmi, DCH 32, DHB 105, Sruthi TCHB 213, NHB 12, HB 224, *etc.*

(*b*) **Intraspecific hybrids.** HH : H 4, H 6, H 8, H 10, NHH 44, JKHY 1, JKHY 2, PKV HY 2, PKVHY 3, PKVHY 4, PKVHY 5, Savita, CICR HH 1, Surya, DHH 11, Fateh, LHH 144, Dhanlaxmi, Maruvikas, Omshankar, CSHH 198, *etc.*

(*c*) **Desi hybrids.** DH 7, DH 9, DDH 2, Pha 46, LDH 11, AKDH 7, RAJDH 7, GCot MDH 11, AAH 1, CICR 2, *etc.*

(*d*) **Male Sterility based hybrids.** Suguna, PKVHY 3, PKVHY 4, PKVHY 4, AAH 1, ARDH 7, RAJDH7, G.Cot MDH 11, CICR 2.

(*e*) **Bt. Hybrids.** About 619 Bt. cotton hybrids have been released by different seed companies.

Nucleus Seed Production

Nucleus seed is the base material in seed production chain. The important steps involved in the nucleus seed production of cotton are briefly presented as follows:

(i) Base Material: Nucleus seed is produced from nucleus seed or breeder seed or sometimes from foundation seed. Select good quality nucleus or breeder or foundation seed for production of nucleus seed. Grow the breeder seed in large plot of at least 0.25 acre. The field of breeder seed production already planted can be used for production of nucleus seed. There is no need of planting separate field for such purpose.

(ii) Breeding Method Used: In cotton, plant to progeny selection is used for production of nucleus seed. Each selected true to the type single plant is selfed. At maturity, bolls of each selected plant are harvested separately and stored in separate envelope.

(iii) Number of Plants to be selected: It depends on the quantity of nucleus seed to be produced. Generally, 200 to 250 single plants true to type of the variety are selected from the base population during flower initiation stage. These single plants are selfed and bolls of each selected single plant are harvested separately.

(iv) Planting Progeny Rows: In the second crop season progeny row of each selected single plant is planted separately adopting 50m isolation distance. Each progeny of single plant is carefully examined from flowering onwards. Off type plants are removed from each progeny row. If variants in a progeny row are many, the entire progeny row is rejected.

(v) Laboratory Evaluation: Each selected single plant is evaluated for fibre properties, *viz.* fibre length, strength, fineness and maturity in the laboratory. Seed shape, size and colour are also examined.

(vi) Seed Bulking: The seed of all true to the type progeny rows is bulked. This constitutes nucleus seed, which is used for production of breeder seed or nucleus seed in the third crop season. The bulk seed is grown in larger plots to produce breeder seed for further use in seed production chain.

TABLE 16.1: Steps in Nucleus Seed Production of Cotton

Sl.No.	*Season*	*Particulars*	*Details*	*Remarks*
1	First	Base Population to be raised	0.25 Acre	In disease free field
		Single plants to be selected	200	Tag selected plants
2	Second	Plant rows of progeny single plants	200	In isolation of 50m
		Laboratory evaluation of plant to row progeny	150	For seed characters
		Mixing true to type progeny rows	100	This constitutes nucleus seed
3	Third	Breeder seed production		Using nucleus seed

Note: Keep some nucleus seed in reserve for use in case of crop failure.

2. Jute

Jute is an important commercial fibre yielding crop of India next to cotton. There are two cultivated species of jute, *viz. Corchorus olitorius* and *Corchorus capsularis.*

Jute is predominantly a self-pollinated crop. Average cross-pollination is 1.38 per cent in *C. capsularis* and 10.5 per cent in *C. olitorius*. Jute is a seed propagated crop. Important jute producing states are West bengal, Andhra Pradesh, Maharashtra, Karnataka and Gujarat.

Nucleus Seed Production

Nucleus seed is the base material in seed production chain. The important steps involved in the nucleus seed production of maize are briefly presented as follows:

(i) **Base Material**: Nucleus seed is produced from nucleus seed or breeder seed or sometimes from foundation seed. Select good quality nucleus or breeder or foundation seed for production of nucleus seed. Grow the breeder seed in large plot of at least 0.25 acre. The field of breeder seed production already planted can be used for production of nucleus seed. There is no need of planting separate field for such purpose.

(ii) **Breeding Method Used**: In Jute, plant to progeny selection is used for production of nucleus seed. Each selected true to the type single plant is selected at maturity and seeds of each selected plant are harvested separately and stored in separate envelope.

(iii) **Number of Plants to be selected**: It depends on the quantity of nucleus seed to be produced. Generally, 500 to 600 single plants true to type of the variety are selected from the base population during flower initiation stage. These single plants are harvested separately.

(iv) **Planting Progeny Rows**: In the second crop season progeny row of each selected single plant is planted separately adopting 50m isolation distance. Each progeny of single plant is carefully examined from flowering onwards. Off type plants are removed from each progeny row. If variants in a progeny row are many, the entire progeny row is rejected.

(v) **Laboratory Evaluation:** Each selected single plant is evaluated for fibre properties, *viz.* fibre length, strength, fineness and maturity in the laboratory. Seed shape, size and colour are also examined.

(vi) **Seed Bulking:** The seed of all true to the type progeny rows is bulked. This constitutes nucleus seed, which is used for production of breeder seed or nucleus seed in the third crop season. The bulk seed is grown in larger plots to produce breeder seed for further use in seed production chain.

Summary

Important commercial Fibre crops include cotton, jute, ramie, *etc.* An area of plant breeding which deals with principles and methods of breeder seed production is referred to as maintenance breeding or varietal maintenance technology. Main points related to base material, breeding procedure used, main focus, plant part used, progenies grown, isolation distance, selection of field, crop inspection and evaluation of selected progenies for maintenance breeding [nucleus seed production] of cotton, jute, ramie, *etc.* are discussed.

QUESTIONS

1. Describe briefly genetic origin, species composition and breeding objectives of Cotton
2. Discuss in brief breeding procedure used for genetic improvement of Cotton crop.
3. Explain in short practical achievements of Cotton breeding in India.
4. Describe in brief important cotton breeding centres and their role in varietal improvement.
5. Describe briefly the procedure of nucleus seed production in upland cotton.
6. Discuss briefly the procedure of nucleus seed production in Jute.

* * * *

Chapter 17

Nucleus Seed Production of A, B and R Lines

Introduction

The nucleus seed production of hybrids involves the nucleus seed production of their parental lines, *viz.* A, B and R lines. In case of pearl millet and *Sorghum* generally single cross hybrids are developed for commercial cultivation. The success of hybrid depends on the maintenance of genetic purity of parental lines. Single cross hybrids in these crops are developed by using nucleus seed of parental lines [A, B and R lines]. A brief account of nucleus seed production of A line [CMS], B line and R line in these two crops is presented as follows:

1. Sorghum (*Sorghum bicolour* Moench)

Sorghum is seed propagated crop. It is a self-pollinated crop. Natural cross pollination varies from 5 to 30 per cent depending upon genotype and environmental conditions. Pollen is very light and cross pollination occurs by wind. In *Sorghum*, protogyny is common which promotes cross pollination. An isolation distance of 200 m. and 100 m. is essential for the production of foundation and certified seed.

Nucleus Seed Production of A and B Line

Nucleus seed is the base material in seed production chain. In *Sorghum*, important steps involved in the nucleus seed production of A, and B lines are briefly presented as follows:

(i) **Base Material**: The nucleus seed of A and B line is produced from nucleus seed or breeder seed. Select good quality nucleus or breeder seed for production of nucleus seed of A and B lines. Grow the breeder seed in large plot of at least 0.25 acre. The field of breeder seed production already planted can be used for production of nucleus seed. There is no need of planting separate field for such purpose.

(ii) **Breeding Method Used**: In *Sorghum*, panicle to row selection is used for production of nucleus seed. One panicle true to the type is selected per plant at panicle emergence and subjected to selfing. Seed of each selected panicle is harvested separately and stored in separate envelope.

(iii) **Number of Panicle to be selected**: It depends on the quantity of nucleus seed to be produced. Generally, 200 to 300 panicles true to type each of A and B lines are selected and harvested separately at maturity. Paired crosses between selected plants of A and B lines are designated such as A1 x B1, A2 x B 2, A3 x B3 ———→ A200 x B200.

(iv) **Planting Progeny Rows**: In the second crop season, progeny row of each selected panicles [200] of A line and A x B crosses is planted separately. An isolation distance of 200m is kept all around the progeny rows. Each panicle to row progeny is carefully examined from panicle emergence onwards for diagnostic characters. Select about 50 best progenies of A and B lines conforming to the characters of parental lines. Off type plants are removed from each progeny row of A and B lines. If variants in a progeny row are many, the entire progeny row is rejected.

(v) **Laboratory Evaluation:** In the laboratory, seed of 50 finally selected progeny rows of A and B lines is evaluated in terms of shape, size, colour and lustre.

(vi) **Seed Bulking:** Finally the seed of 50 true to the type progeny rows of A and B lines is bulked. This constitutes nucleus seed of A and B lines, which is used for production of single cross hybrid in third crop season. The bulk seed is grown in larger plots to produce single cross hybrid.

TABLE 17.1: Steps in Nucleus Seed Production of A and B Lines in Sorghum

Sl.No.	*Season*	*Particulars*	*Details*	*Remarks*
1	First	Base Population to be raised	0.25 Acre	In disease free field
		Panicles to be selected and selfed Panicles finally to be selected	1000 200	Tag selected ears
2	Second	Plant panicle to rows progeny	200	In isolation of 200m
		Laboratory evaluation of panicle row progeny	200	For seed characters
		Bulking equal number of seed from panicles	150	This constitutes nucleus seed
3	Third	Breeder seed production		Using nucleus seed

Note: Keep some nucleus seed in reserve for use in case of crop failure.

Nucleus Seed Production of R Line

Restorer line is used for restoration of fertility in hybrid developed for commercial seed production. The success of hybrid also depends on the genetic purity of R line used in single cross hybrid. Important steps involved in the production of nucleus seed of R line in Sorghum are briefly presented as follows:

(i) **Base Material**: Nucleus seed of R line is produced from nucleus seed or breeder seed. Select good quality nucleus or breeder seed for production of nucleus seed of R line. Grow the breeder seed in large plot of at least 0.25 acre. The field of breeder seed production already planted can be used for production of nucleus seed. There is no need of planting separate field for such purpose.

(ii) **Breeding Method Used**: In *Sorghum*, panicle to row selection is used for production of nucleus seed. One panicle true to the type is selected per plant at panicle emergence and subjected to selfing. Seed of each selected panicle is harvested separately and stored in separate envelope.

(iii) **Number of Panicle to be selected**: It depends on the quantity of nucleus seed to be produced. Generally, 200 to 300 panicles true to type of R line are selected and harvested separately at maturity.

(iv) **Planting Progeny Rows**: In the second crop season, progeny row of each selected panicle of R line is planted separately. An isolation distance of 200m is kept all around the progeny rows. Each panicle to row progeny is carefully examined from panicle emergence onwards for diagnostic characters. Off type plants are removed from each progeny row of R line. If variants in a progeny row are many, the entire progeny row is rejected.

(v) **Laboratory Evaluation:** In the laboratory, seed of all finally selected progeny rows of R line are evaluated in terms of shape, size, colour and lustre.

(vi) **Seed Bulking:** Finally the seed of 200 true to the type progeny rows of R line is bulked. This constitutes nucleus seed of R line, which is used for production of single cross hybrid in third crop season. The bulk seed is grown in larger plots to produce single cross hybrid.

TABLE 17.2: Steps in Nucleus Seed Production of R Line in Sorghum

Sl.No.	*Season*	*Particulars*	*Details*	*Remarks*

Sl.No.	*Season*	*Particulars*	*Details*	*Remarks*
1	First	Base Population to be raised	0.25 Acre	In disease free field
		Panicles to be selected and selfed Panicles finally to be selected	1000 200	Tag selected ears
2	Second	Plant panicle to rows progeny	200	In isolation of 200m
		Laboratory evaluation of panicle row progeny	200	For seed characters
		Bulking equal number of seed from panicles	150	This constitutes nucleus seed
3	Third	Breeder seed production		Using nucleus seed

Note: Keep some nucleus seed in reserve for use in case of crop failure.

2. Pearl Millet

Pearl millet is a seed propagated and cross-pollinated crop. In this crop natural cross pollination is more than 80 per cent. Pollen is very light and cross pollination occurs by wind. In pearl millet protogyny is common which promotes cross pollination. An isolation distance of 400 m. and 200 m is essential for the production of foundation and certified seed.

Nucleus Seed Production of A and B Line

Nucleus seed is the base material in seed production chain. In pearl millet, important steps involved in the nucleus seed production of A, and B lines are briefly presented as follows:

(i) **Base Material**: The nucleus seed of A and B line is produced from nucleus seed or breeder seed. Select good quality nucleus or breeder seed for production of nucleus seed of A and B lines. Grow the breeder seed in large plot of at least 0.25 acre. The field of breeder seed production already planted can be used for production of nucleus seed. There is no need of planting separate field for such purpose.

(ii) **Breeding Method Used**: In pearl millet, ear to row selection is used for production of nucleus seed. One ear true to the type is selected per plant at ear emergence and subjected to selfing. Seed of each selected ear is harvested separately and stored in separate envelope.

(iii) **Number of Ears to be selected**: It depends on the quantity of nucleus seed to be produced. Generally, 200 to 300 ear true to type each of A and B lines are selected and harvested separately at maturity. Paired crosses between selected plants of A and B lines are made such as A1 x B1, A2 x B 2, A3 x B3 ———→ A200 x B200.

(iv) **Planting Progeny Rows**: In the second crop season, progeny row of each selected A line and A x B crosses is planted separately. An isolation distance of 400m is kept all around the progeny rows. Each ear to row progeny is carefully examined from ear emergence onwards for diagnostic characters. Select about 50 best progenies each of A and B lines conforming to the characters of parental lines. Off type plants are removed from each progeny row of A and B lines. If variants in a progeny row are many, the entire progeny row is rejected.

(v) **Laboratory Evaluation:** In the laboratory, seed of 50 finally selected progeny rows of A and B lines are evaluated in terms of shape, size, colour and lustre.

(vi) **Seed Bulking:** Finally the seed of 50 true to the type progeny rows of A and B lines is bulked. This constitutes nucleus seed of A and B lines, which is used for production of single cross hybrid in third crop season. The bulk seed is grown in larger plots to produce single cross hybrid.

TABLE 17.3: Steps in Nucleus Seed Production of A and B Lines in Pearl Millet

Sl.No.	*Season*	*Particulars*	*Details*	*Remarks*
1	First	Base Population to be raised	0.25 Acre	Self all plants
		Single plants to be selected	200	Selfed Plants
2	Second	Progeny rows of single plants	200	In isolation
		Laboratory evaluation of single plant progeny	100	For seed characters
		Mixing true to type progeny rows	75	This constitutes nucleus seed
3	Third	Breeder seed production		Using nucleus seed

Note: Keep some nucleus seed in reserve for use in case of crop failure.

Nucleus Seed Production of R Line

Restorer line is used for restoration of fertility in hybrid developed for commercial seed production. The success of hybrid also depends on the genetic purity of R line used in single cross hybrid. Important steps involved in the production of nucleus seed of R line in Pearl millet are briefly presented as follows:

(i) **Base Material**: Nucleus seed of R line is produced from nucleus seed or breeder seed. Select good quality nucleus or breeder seed for production of nucleus seed of R line. Grow the breeder seed in large plot of at least 0.25 acre. The field of breeder seed production already planted can be used for production of nucleus seed. There is no need of planting separate field for such purpose.

(ii) **Breeding Method Used**: In Pearl millet, ear to row selection is used for production of nucleus seed. One ear true to the type is selected per plant at ear emergence and subjected to selfing. Seed of each selected ears is harvested separately and stored in separate envelope.

(iii) **Number of Ears to be selected**: It depends on the quantity of nucleus seed to be produced. Self about 1000 plants true to type of R line finally select about 200 ears and harvested separately at maturity.

(iv) **Planting Progeny Rows**: In the second crop season, progeny row of each selected ears of R line is planted separately. An isolation distance of 400m is kept all around the progeny rows. Each ear to row progeny is carefully examined from ear emergence onwards for diagnostic characters. Off type plants are removed from each progeny row of R line. If variants in a progeny row are many, the entire progeny row is rejected.

(v) **Laboratory Evaluation:** In the laboratory, seed of all finally selected progeny rows of R line are evaluated in terms of shape, size, colour and lustre.

(vi) **Seed Bulking:** Finally the seed of 200 true to the type progeny rows of R line is bulked. This constitutes nucleus seed of R line, which is used for production of single cross hybrid in third crop season. The bulk seed is grown in larger plots to produce single cross hybrid.

TABLE 17.4: Steps in Nucleus Seed Production of R Line in Pearl Millet

Sl.No.	*Season*	*Particulars*	*Details*	*Remarks*
1	First	Base Population to be raised	0.25 Acre	In disease free field
		Panicles to be selected and selfed Panicles finally to be selected	1000 200	Tag selected ears
2	Second	Plant panicle to rows progeny	200	In isolation of 200m
		Laboratory evaluation of Ear row progeny	200	For seed characters
		Bulking equal number of seed from panicles	150	This constitutes nucleus seed
3	Third	Breeder seed production		Using nucleus seed

Summary

The nucleus seed production of hybrids involves the nucleus seed production of their parental lines, *viz.* A, B and R lines. In case of pearl millet and Sorghum generally single cross hybrids are developed for commercial cultivation. The success of hybrid depends on the maintenance of genetic purity of parental lines. Single cross hybrids in these crops are developed by using nucleus seed of parental lines [A, B and R lines]. A brief account of nucleus seed production of A line [CMS], B line and R line in these two crops is outlined

Glossary

A Line: A male sterile line; It may be genetic, cytoplasmic and genic cytoplasmic.

B Line: An isogenic line of A line which is male fertile.

Cytoplasmic Male Sterility: Male sterility governed by cytoplasmic genes found either in chloroplasts bot mitochondria.

Gene Cytoplasmic Male Sterility: Male sterility controlled by the combination of both nuclear and cytoplasmic genes.

Genetic Male Sterility: Male sterility controlled by nuclear genes.

Isogenic Line: A line with single gene difference.

R Line: A line which has fertility restoration capacity in F1 hybrids.

QUESTIONS

1. **Describe briefly the procedure of nucleus seed production of A and B lines in Sorghum.**
2. **Explain briefly the procedure of nucleus seed production of R line in Sorghum.**
3. **Discuss briefly the procedure of nucleus seed production of A and B lines in pearl millet.**

4. **Describe briefly the procedure of nucleus seed production of R line in Pearl millet.**

5. **Write Short Notes on the following:**

   (i) A Line

   (ii) B Line

   (iii) R Line

   (iv) Isolation Distance

* * * *

Chapter 18

Intellectual Property Rights

Introduction

The property refers to wealth or valuable things earned by a person. It is estimated in terms of land, house, garden, industries, animals, gold, silver, diamond, money, *etc.* accumulated by a person. Each person wants to earn property for four main purposes as given below:

1. To get a better or higher status in the society, because status of a person is measured in terms of property accumulated by him or her. On the basis of property, the person is called as rich, middle class or poor.
2. To lead a better quality of life in the society.
3. To get a name and fame in the society. In the society, rich people get more respect than others.
4. For better financial security of life for himself and for his children.

Types of Property

The various valuable belongings of a person are broadly classified into two groups, *viz.*, 1. immovable property, and 2. movable property. These are defined below:

1. **Immovable Property:** It refers to fixed types of properties which cannot move from one place to other. Such properties include land, buildings [house, flat, bungalow, farm house] and gardens. Such properties are protected by laws of land in each country.
2. **Movable Property:** It refers to the property which can be easily shifted from one place to other. Such property includes animals, farm machines, furniture, fixtures, gold, silver, diamond, and money. Such properties are also protected by public laws.

3. **Intellectual Property:** There is third type of property what we call intellectual property. The product/process/idea which is outcome of the brain of a person and can be used on commercial scale for benefit of human kind is called intellectual property. In other words, Intellectual property refers to creations of the mind: inventions, literary and artistic works, and symbols, names, images, and designs used on commercial scale. The main features of intellectual property are given below:
   (a) It is measured in terms of new ideas, processes, products, inventions and innovations developed by a person.
   (b) It requires lot of intellectual inputs in terms of thinking, planning and fine tuning of new ideas/products/processes, *etc.*
   (c) It requires considerable amount of funds and other resources to develop new products/processes.
   (d) The main problem with intellectual property is that it can be copied, reproduced and used by others resulting in loss of inventor. Hence protection of intellectual property is essential so that the inventor can derive maximum benefits from his invention.

The rights to intangible property that is the product of the human intellect are referred to as intellectual property rights. Intellectual property may be protected by copyright, trademark or patent. The holder of intellectual property rights is usually the person or persons who developed the product or the organization that funded it.

Forms of Intellectual Property Rights

The Intellectual Property Rights are broadly divided into two groups, *viz.*, Primary Rights and Sui-generis Rights. These are briefly discussed below:

A. **Primary Rights:** Primary rights include: Copyrights, Patents, Trademarks, Trade Secret, Trade names, Domain names, Industrial Designs, Geographical Indications, *etc.*

TABLE 18.1: Forms of Intellectual Property Rights/Laws

Sl.No.	*Primary Rights*	*Sl.No.*	*Sui Generis Rights*
1	Copyrights	1	Database Rights
2	Patents	2	Mask work
3	Trademarks	3	Plant Breeders'Rights
4	Industrial Design Rights	4	Farmers'Rights
5	Utility Models	5	Moral Rights
6	Geographical Indication	6	Supplementary Protection Certificate
7	Trade Secret	7	Indigenous Intellectual Property
8	Related Rights		
9	Trade Names		
10	Domain Names		

B. Sui-generis Rights: Sui-generis refers to things of their own kind or things with unique characteristics. Such rights include Database Rights, Mask work, Plant Breeders Rights, Traditional Knowledge, Moral Rights and Supplementary Protection Certificate.

A. Primary Rights

1. Copyrights

The legal right granted to an author, composer, playwright, publisher, or distributor to exclusive publication, production, sale, or distribution of a literary, musical, dramatic, or artistic work is called copyright. Main features of copyrights are briefly presented below:

(i) It provides protection for a specific duration.

(ii) It is applicable in all countries.

(iii) The copyright holder has the rights to authorize others to use the protected work.

(iv) It is applicable to books, movies, music, paintings, photographs and software.

(v) The information cannot be reproduced as such without written permission. However, the information or idea can be used by any one.

2. Patent

Patent refers to a document granting an inventor sole rights to an invention. It is an official document which grants sole rights to the inventor for manufacturing and marketing his product/process/invention to derive benefits. In other words, patent is a set of exclusive rights granted by a state to a patentee (the inventor or assignee) for a fixed period of time. Main points are as follows.

1. Patents grant the right to prevent others from copying the invention.
2. An official license is granted by the Patent Office to issue exclusive right to an individual or business for production or sale of a specific invention.
3. It is granted by a government to an inventor, assuring the sole right to make, use, and sell the invention for a certain period of time. Many audio and video technologies are covered by patents.
4. It is a document that allows the patent owner to prevent others from making, using or selling the invention protected by the patent.

3. Trademark

A trademark or trade mark is a distinctive sign of some kind which is used by an individual, business organization or other legal entity to uniquely identify the source of its products and/or services to consumers, and to distinguish its products or services from those of other entities. Main features of trademarks are briefly presented below:

1. A trademark can be a word, name, symbol, device or mark which is used to identify and distinguish the goods or services of one company from goods or services of another.
2. Trademark is used to identify its product and to distinguish them from others. It is the name of a product made by a particular person or company.
3. The period of protection for a trademark varies, but can generally be renewed indefinitely.

4. Industrial Design

An industrial design - or simply a design - is the ornamental or aesthetic aspect of an article produced by industry or handicraft; registration and renewals provide protection for, in most cases, up to 15 years. Main points about industrial designs are given below:

1. Industrial design is an applied art whereby the aesthetics and usability of products may be improved for marketability and production.
2. Industrial design refers to any original shape, picture, or some combination applied to a useful article of manufacture.
3. Industrial design refers to the professional service of creating and developing concepts and specifications that optimize the function, value, and appearance of products and systems for the mutual benefit of both user and manufacturer [Industrial Design Society of America].

5. Utility Models

A utility model is an exclusive right granted for an invention, which allows the right holder to prevent others from commercially using the protected invention, without his authorization, for a limited period of time. A utility model is similar to a patent. In fact, utility models are also referred to as "petty patents" or "innovation patents, minor patent and small patent." The important points about utility model are listed below:

1. Utility model is an intellectual property rights to protect inventions.
2. Utility models are also known as petty patents, petty inventions, petty innovations, utility innovations, minor patents and small patent.
3. The rights conferred by utility models are similar to those granted by patent laws but has a shorter term.
4. They are registered with National Patent Office.
5. They are more suited to incremental inventions.

6. Geographical Indications

A geographical indication is a sign used on goods that have a specific geographical origin and often possess qualities or a reputation that are due to that place of origin. In other words, a geographical indication is a name or sign used on certain products or which corresponds to a specific geographical location or origin (*e.g.* a town, region, or country). Main points are as follows:

1. This is the name of a region, a specific place or, in exceptional cases a country, used to describe an agricultural, natural or manufactured goods product or a food stuffs:
2. So far geographical indications have been registered in some advanced countries because they have developed the system of protecting GI.
3. In the past, wines, spirits, cheeses, tobacco, which account for 88 per cent, have been registered as GI. Wines and spirits account for almost 71 per cent of all registrations.
4. In addition to wines and spirits, there is need for protection of agricultural products as GI. For this there is need to establish a multilateral system of identification and registration of GI.
5. Geographical Indication may be an agricultural, natural or manufactured goods or product.

7. Trade Secret

A "trade secret" is anything [a formula, process, method, mechanism, tool, pattern or device] which the disclosing party desires to keep secret. Trade secrets usually include such things as the manufacturing details for a product, variations or alternative uses.

The main points about trade secret are briefly presented as follows:

(i) There is no specific period for trade secret. It may continue lifelong or for generations together.
(ii) There is no need of registration for trade secret.
(iii) It does not provide opportunity to others for improvement of innovation.
(iv) It is not applicable to books, equipment, plant varieties, designs which are openly used.

8. Related Rights

Related rights is a term in copyright law, used in opposition to the term "authors' rights". The term neighboring rights is exactly equivalent to related rights. Both authors' rights and related rights are copyrights in the sense of English or U.S. law. Main points about related rights are given below:

1. Related rights are similar to copyrights.
2. Related rights come under primary property rights.
3. Related rights are also known as neighboring rights.
4. Related rights are closer to copyrights but are not covered by the Berne Convention.
5. Related Rights are covered by Rome Convention, 1961.

9. Trade Names

A **trade name**, also known as a trading name or a business name, is the name which a business trades under for commercial purposes. Main points related to trade name are given below:

1. It is also known as business name, trading name, assumed name, brand name and corporate name.
2. It may or may not be registered
3. It distinguishes a particular business from others.
4. It can be exclusive or non-exclusive.
5. It is generally used on letter heads and bank accounts.
6. It is the name given to a particular substance by each company that manufactures it.
7. It is the name under which a company conducts its business.
8. It is the business name of the person or organization making and/or selling a particular product. "General Motors, Inc" is the "trade name" of the company
9. Trade names are registered by the state in which the person or company is based.

10. Domain Names

Domain refers to the unique name that identifies an Internet site. Domain names have two or more parts separated by dots. For example www.ask-edi.com or www.bcbsks.com The term domain name has multiple related meanings:

1. It identifies a web site on the internet (eg, company.com); also referred to as a URL.[Uniform Resource Locator]
2. It allows reference to Internet sites without knowing the true numerical address.
3. It refers to the official name of a computer connected to the Internet. (eg, "www.microsoft.com").
4. It is the unique name of a computer on the Internet that distinguishes it from other systems around the world.
5. They are sometimes colloquially (and incorrectly) referred to by marketers as "web addresses".

B. Sui-generis Rights

Sui-generis is a Latin expression, literally meaning of its own kind/genus or unique in its characteristics. In Intellectual Property there are rights which are known as being sui-generic to owners of a small class of works. Such Intellectual Property Rights include (i) database Rights, (ii) mask work, (iii) plant breeders' rights, (iv) farmers' rights, (v) moral rights, (vi) supplementary protection certificate, and (vii) indigenous intellectual property, A brief account of all these is presented in this chapter.

1. Database Rights

Database rights are a form of exclusive right introduced by European Union Law in 1996 in those countries which follow EU Law. A 'database right' is an intellectual property right given to a computer database. The main points about database rights are as follows:

(i) In most countries database are covered by copyright law to some degree.

(ii) The database rights were introduced by European Union Law to provide greater protection to collection of information.

(iii) Database right is covered by sui generic rights.

(iv) Duration: Database rights last for 15 years, but can be extended if the database is updated.

(v) Protection: Database right prevents copying of substantial parts of database. The protection is over the information, not over the form of expression.

TABLE 18.2: List of Sui-generis Rights

Sl.No.	*Items of Sui-generic Right*	*Definition or Brief Description*
1.	Database rights	A 'database right' is an intellectual property right given to a computer database
2.	Mask Work	A mask work is a two or three-dimensional layout of an integrated circuit (IC or "chip")
3.	Plant Breeders' Rights	Rights granted to the breeder of a new variety of plant
4	Farmers' Rights	The legal rights provided to farmers to save, use, sow, replant, exchange, share or sell his farm produce
5	Moral Rights	Moral rights are a special extension to copyright that gives the originator of a copyrighted work rights over its use.
6.	Supplementary protection certificate	A supplementary protection certificate (SPC) is a sui generis, patent-like, intellectual property right that is available for medicinal and plant protection products.
7.	Indigenous Intellectual Property	The knowledge, innovations and practices of indigenous people and local communities.

2. Mask Work Rights

A mask work is a two or three-dimensional layout of an integrated circuit (IC or "chip"), *i.e.* the arrangement on a chip of semiconductor devices such as transistors and passive electronic components such as resistors and interconnections. Main points related to mask work are listed as under.

1. Mask work cannot be protected under copyright law or patent law.
2. The mask work exclusive rights were first granted in USA by Semiconductor Chip Protection Act of 1984. Such legislations also exist in Canada and Australia.
3. **Duration.** Rights for semiconductor mask work lasts only for two years (if unregistered) or 10 years (if registered).

4. The exclusive rights in mask work are somewhat like those of copyright: the right to reproduce the mask work or distribute an IC made using mask work.
5. Reproduction for reverse engineering of a mask work is permitted by law.
6. Mask work rights exist when they are created, even without registration
7. Protection under mask work cannot be provided to a mask that is not original.

Database and Copyright

There are some similarity and some differences between database and copyrights which are presented in Table 18.3.

TABLE 18.3: Comparison of Database and Copyrights

Sl.No.	*Particulars*	*Database Rights*	*Copyrights*
1	Governed by	Sui generic rights	Primary intellectual rights
2.	Registration	Not compulsory	Not compulsory
3	Period of protection	15 years	70-120 years
4.	Protection over	Information	Expression
5.	Validity	Country of registration	Country of registration
6.	Enforcement	After creation	After creation

Database and Mask Work

There are some similarities and some differences between database and mask work which are presented in Table 18-4.

TABLE 18.4: Comparison of database and Mask work

Sl.No.	*Particulars*	*Database Rights*	*Mask Work*
1	Governed by	Sui generic rights	Sui generic rights
2.	Registration	Not compulsory	Not compulsory
3	Period of protection	15 years	Unregistered 2 Registered 10 years
4.	Protection over	Information	Integrated Circuits
5.	Validity	Country of registration	Country of registration
6.	Enforcement	After creation	After creation

3. Plant Breeders' Rights

Plant breeders' rights, also known as plant variety rights (PVR), are intellectual property rights granted to the breeder of a new variety of plant. Plant Breeders Rights are granted to novel plant varieties that are distinctive, uniform, and stable (*e.g.*, cultivars bred true-to-type for desired traits). The legal protection of a new plant variety is granted to the breeder or his successor. The effect of PBR is that prior authorization is required before the material can be used for commercial purposes.

Main Features

Main features of Plant Breeders' Rights in relation to registration, duration, validity, matters covered, requirements, transfer, control, enforcement, rights protected and infringement are briefly presented below:

1. **Registration:** Registration is essential to get legal rights. There is no protection without registration. A nominal processing Fee is charged for registration.
2. **Duration**: The period of protection varies with plant species. It is 20 years for field crops, 25 years for trees, vines, fruit trees, ornamental trees and shrubs [UPOV 1991].
3. **Validity:** The protection of the Plant Breeders' Rights is valid only in the country where it has been registered. The protection in other countries can be obtained by filling separate application in each country.
4. **Matters Covered**: The protection right can be granted for varieties of all botanical genera and species. The variety should have a designation [name] as per the rule of International Code of Nomenclature
5. **Requirements:** There are four basic requirements for protection of a variety under PBR, *viz.* (i) novelty, (ii) distinctiveness, (iii) uniformity and (iv) stability.
6. **Transfer:** An authorized plant breeder has right to authorize other interested persons for commercial production and marketing of his variety on royalty basis.
7. **Control:** The authorized plant breeder has right to prevent other from commercial production and marketing of his variety without permission
8. The Plant Breeders' Rights come into force immediately after registration of the variety.
9. **Rights Protected**: It provides exclusive rights to the owner for commercial production and marketing of his variety.
10. **Infringement.** Unauthorised production and marketing of a registered variety by other person amounts to infringement. The owner has the right to take legal action against the infringer and claim damages.

Advantages of PBR

1. Breeders get benefit of their variety.
2. PBR help in faster development of seed industry.
3. PBR lead to improvement in quality because of competition.
4. PBR are useful in procurement of good material on payment basis.
5. PBR help in enrichment of genetic resources.

Disadvantages of PBR

1. It will promote monopoly.

2. It will encourage unhealthy practices.
3. It may lead to increase in prices.
4. There will be reduction in genetic variability.
5. There will be compulsion to purchase fresh seed every year.

4. Farmers' Rights

Farmers' rights refer to the rights arising from the past, present and future contributions of farmers in conserving, improving and making available plant or animal genetic resources, particularly those in the centres of origin/diversity. In other words, the legal rights provided to farmers to save, use, sow, replant, exchange, share or sell his farm produce including seed of a variety protected under Plant Variety Protection Act refer to Farmers Rights. The purpose of these rights is to "ensure full benefits to farmers and support the continuation of their contributions

The FAO Conference held in Rome from 11-29 November, 1989 endorsed the concept of Farmers Rights with a view to:

(i) Ensuring global recognition of the need for conservation and the availability of sufficient funds for these purposes;

(ii) Assisting farmers and farming communities throughout the world, especially those in areas of original diversity of plant genetic resources, in the protection and conservation of their PGR and of the natural biosphere; and,

(iii) Allowing the full participation of farmers, their communities and countries in the benefits derived, at present and in the future, from the improved use of PGR.

5. Moral Rights

Moral rights are rights of creators of copyrighted works generally recognized in civil law jurisdictions and first recognized in France and Germany, before they were included in the Berne Convention for the Protection of Literary and Artistic Works in 1928. Moral rights are a special extension to copyright that gives the originator of a copyrighted work rights over its use. The originator can specify moral conditions on the use or exploitation of their creation, even when the rights are licensed or sold to others'. The main points related to moral rights are presented below:

1. These are rights provided to authors and directors of copyrights to protect their personal interest in relation to their creations. These rights are provided beyond those strictly recognized in copyright law.
2. Moral Rights were granted to authors under the Berne Convention (1989) giving the author (i) right to attribution, and (ii)) right to integrity.
3. The creator of the work has the right of attribution and integrity, even after sale or transfer of the copyright. In other words, the creator or author shall have the right to claim authorship of the work even after sale or transfer of the copyright.

4. **Transfer**. The copyrights can be transferred or assigned to others, but moral Right cannot be transferred or assigned.
5. The author has right to object to any distortion, mutilation or other modification which would be prejudicial to his honour or reputation

6. Supplementary Protection Certificate

This term is used in European Countries. This is an extension of the term given to pharmaceutical or plant protection patent. A supplementary protection certificate (SPC) is a sui generis, patent-like, intellectual property right. This type of right is available for medicinal products, such as drugs, and plant protection products, such as insecticides, and herbicides. Main points related to supplementary protection certificate are given below:

1. It is in use in European Union member countries.
2. It is a sui generis, patent like, intellectual property right.
3. The SPC comes into force only after the expiry of corresponding patent.
4. **Duration**. It has maximum life time of 5 years when granted after 10 years after filing date of a patent. The term would be less than 5 years if it is granted after 5 years from the date of filing the corresponding patent. However, the market exclusivity cannot exceed 15 years.
5. The SPC provides long term authorisation to a patent for marketing.
6. **Conditions of Issue.** It is issued against a patent generally immediately on expiry of the term of the patent. It cannot be issued before 5 years from filing date of corresponding patent
7. The registration is required which is issued on national basis. A separately application is to be filed for registration in each country.

7. Indigenous Intellectual Property

The Indigenous Knowledge also known as Traditional Knowledge [TK] refers to the knowledge, innovations and practices of indigenous people and local communities. In other words, it is the: knowledge gained through tradition from ancestors about plants and animals, medicines, agricultural practices and other matters. Main points about traditional knowledge are listed below:

1. Traditional knowledge can be protected by sui generic rights
2. Traditional knowledge is also known as indigenous knowledge or local knowledge or traditional ecological knowledge.
3. It covers the matured long-standing traditions and practices of certain regional, indigenous, or local communities. In other words, it is gained from experience gained over the centuries and adapted to local culture and environment.
4. It includes knowledge about plants, animals, indigenous treatment of diseases through herbal medicines, cultural practices, *etc.*

5. Traditional knowledge is based on teachings and experiences and is orally passed on from generation to generation to some persons only. Earlier there were no written documents on traditional knowledge.
6. Some forms of traditional knowledge are expressed through stories, legends, folklore, rituals, songs and even laws.
7. There is difference between local knowledge and traditional knowledge. The former existed from decade to century, whereas the latter existed from millennia.
8. It is collectively owned by the people or natives of an ecosystem or region.
9. Traditional knowledge incorporates knowledge of ecosystem and code of ethics governing appropriate use of environment. This code includes rules and conventions promoting desirable ecosystem relations, human-animal interactions and even social relationships. A combination of traditional and non-traditional knowledge will form a rich and distinctive understanding of life and the world

The ecological knowledge is composed of both traditional knowledge and experiential knowledge [gained through personal experience

Traditional knowledge is very important almost in every sphere of life. Hence, its protection is very necessary otherwise this valuable knowledge will be gradually lost

Advantages of IPR

Intellectual Property Rights refer to the legal rights provided to an inventor to derive economic benefits from his invention/innovation. The main advantages of IPR are as follows:

1. It promotes healthy competition for invention/innovation among the intellectuals.
2. It helps in improving the quality of the product.
3. It makes available new ideas/technologies to different countries.
4. It leads to faster development of industries/organizations engaged in research and development work.

Disadvantages of IPR

There are some disadvantages of Intellectual Property Rights which are briefly presented below.

1. The procedure of registration, particularly of patents, is very lengthy.
2. It involves lot of money transaction in registration, renewal and licensing.
3. It invites lots of court cases due to infringements
4. It may lead to monopoly of right holders, *etc.*

Summary

The property refers to wealth or valuable things earned by a person. It is estimated in terms of land, house, garden, industries, animals, gold, silver, diamond, money, *etc.* accumulated by a person. The various valuable belongings of a person are broadly classified into two groups, *viz.*, 1. immovable property, and 2. movable property. There is third type of property what we call intellectual property. The product/process/idea which is outcome of the brain of a person and can be used on commercial scale for benefit of human kind is called intellectual property. In other words, Intellectual property refers to creations of the mind: inventions, literary and artistic works, and symbols, names, images, and designs used on commercial scale.

The Intellectual Property Rights are broadly divided into two groups, *viz.*, Primary Rights and Sui-generis Rights. These are listed below:

(i) Primary Rights: Primary rights include: Copyrights, Patents, Trademarks, Trade Secret, Trade names, Domain names, Industrial Designs, Geographical Indications, *etc.*

(ii) Sui-generis Rights: Sui-generis refers to things of their own kind or things with unique characteristics. Such rights include Database Rights, Mask work, Plant Breeders Rights, Traditional Knowledge, Moral Rights and Supplementary Protection Certificate. Details of all these types of intellectual property rights have been presented.

QUESTIONS

1. **Define property. Describe briefly different type of properties**

2. **What do the following terms signify?**

   (i) Immovable property (ii) Movable property

   (iii) Intellectual property (iv) Sui-generis rights

3. **Define the following terms:**

   (i) Copyright (ii) Patent

   (iii) Trademarks (iv) Trade Secret

   (v) Industrial design (vi) Geographical Indication

4. **Differentiate between the following:**

   (i) Movable and immovable property

   (ii) Copyright and patent

   (iii) Trademark and trade Secret

   (iv) Trade Name and Domain Name

5. **What is Intellectual Property? List different forms of Intellectual Properties and describe any one in detail.**

6. **Define exclusive rights. Explain briefly various types of primary rights**

7. **Describe briefly advantages and disadvantages of Intellectual Property Rights.**

8. **Describe briefly the merits and demerits of copyrights, patents, trademark and trade secret.**

9. **Write short notes on any two of the following:**

   (i) Utility Models (ii) Related Rights

   (iii) Trade Name (iv) Domain Name

10. **Give in brief comparison of copyrights and patents.**

11. **Describe briefly trademarks and trade secret. Give their merits and demerits.**

12. **Describe briefly utility models and related rights.**

13. **Describe briefly trade names and domain names.**

14. **What is sui-generic right? List various items covered by such rights and describe any one of them in detail.**

15. **Define database and describe it significance.**

16. **Explain briefly disadvantages of sui generic rights.**

17. **Write short notes on the following:**

   (i) Database rights (ii) Moral Rights

   (iii) Mask Work (iv) Traditional knowledge

18. **Define mask work and discuss its main features.**

19. **What do you mean by plant breeders' rights? Give merits and demerits of PBR.**

20. **Define traditional knowledge. Give its advantages and disadvantages.**

21. **Give a brief account of supplementary protection certificate.**

22. **What do the following terms signify?**

   (i) Sui generic rights (ii) Plant breeders' rights

   (iii) Farmers' rights (iv) Biodiversity

23. **Give a brief comparison of moral rights and traditional knowledge.**

24. **Describe briefly Farmers' Rights and their significance.**

* * * *

Appendices

Appendix 1
List of Agricultural Universities in India

I. AGRICULTURAL UNIVERSITIES

Sl.No.	*Name of University*	*Location*	*State*
A	**STATE UNIVERSITIES**		
1	Acharya NG Ranga Agricultural University	Hyderabad	Telangana
2	Assam Agricultural University.	Jorhat	Assam
3	Bihar Agricultural University.	Sabour	Bihar
4	Birsa Agricultural University	Ranchi	Jharkhand
5	Indira Gandhi Krishi Vishwavidyalaya.	Raipur	Chhatishgarh
6	Anand Agricultural University.	Anand	Gujarat
7	Junagadh Agricultural University.	Junagarh	Gujarat
8	Navsari Agricultural University.	Navsari	Gujarat
9	Sardarkrushinagar-Dantiwada Agricultural University.	Bansakantha	Gujarat
10	Chaudhary Charan Singh Haryana Agricultural University.	Hisar	Haryana
11	CSK Himachal Pradesh Krishi Vishvavidyalaya,	Palampur	Himachal Pradesh
12	Sher-E-Kashmir Univ of Agricultural Sciences and Technology.	Jammu	Jammu and Kashmir
13	Sher-E-Kashmir Univ of Agricultural Sciences and Technology of Kashmir.	Srinagar	Jammu and Kashmir
14	University of Agricultural Sciences.	Bangalore	Karnataka
15	University of Agricultural Sciences.	Dharwad	Karnataka
16	University of Agricultural Sciences.	Raichur	Karnataka
17	Kerala Agricultural University.	Vallanikara	Kerela
18	Jawaharlal Nehru Krishi Viswavidyalaya.	Jabalpur	Madhya Pradesh
19	Rajmata Vijayaraje Scindia Krishi Viswavidyalaya.	Gwalior	Madhya Pradesh
20	Dr. Balasaheb Sawant Konkan Krishi Vidyapeeth.	Ratnagiri	Maharashtra
21	Dr. Panjabrao Deshmukh Krishi Vidyapeeth.	Akola	Maharashtra
22	Mahatma Phule Krishi Vidyapeeth.	Rahuri	Maharashtra
23	Marathwada Agricultural University.	Parbhani	Maharashtra
24	Odisha Univ. of Agriculture and Technology.	Bhubaneswar	Odisha
25	Punjab Agricultural University.	Ludhiana	Punjab
26	Maharana Pratap Univ. of Agriculture and Technology.	Udaipur	Rajasthan
27	Rajasthan Agricultural University	Bikaner	Rajasthan
28	Agriculture University	Kota	Rajasthan
29	Sri Karan Narendra Agriculture University	Jobner	Rajasthan
30	Agriculture University	Jodhpur	Rajasthan
31	Tamil Nadu Agricultural University.	Coimbatore	Tamil Nadu

Sl.No.	Name of University	Location	State
32	Chandra Shekar Azad University of Agriculture and Technology.	Kanpur	Uttar Pradesh
33	Narendra Deva University of Agriculture and Technology.	Faizabad	Uttar Pradesh
34	Sardar Ballabh Bhai Patel Univ. of Agriculture and Technology.	Meerut	Uttar Pradesh
35	Manyavar Kashiram University of Agriculture and Technology.	Banda	Uttar Pradesh
36	Govind Ballabh Pant University of Agriculture and Technology.	Pantnagar	Uttrakhand
37	Bidhan Chandra Krishi Viswavidyalaya.	Kalyani	West Bengal
38	Uttar Banga Krishi Viswavidyalaya.	Cooch Bihar	West Bengal
B	**CENTRAL UNIVERSITIES**		
1	Central Agricultural University	Imphal	Manipur
2	Rani Laxmibai Central Agricultural University	Jhansi	Uttar Pradesh
3	Rajendra Central Agricultural University.	Pusa	Bihar
C	**DEEMED UNIVERSITIES**		
1	Indian Agricultural Research Institute.	Pusa	New Delhi
2	National Dairy Research Institute	Karnal	Haryana
3	Indian Veterinary Research Institute	Bareilly	Uttar Pradesh
4	Central Institute for Fisheries Education	Mumbai	Maharashtra
5	Allahabad Agricultural Institute.	Allahabad	Uttar Pradesh

II. ANIMAL SCIENCE UNIVERSITIES

Sl.No.	Name of University	Location	State
1	Sri Venkateswara Veterinary University.	Tirupati	Andhra Pradesh
2	Karnataka Veterinary, Animal and Fisheries Sciences University.	Bidar	Karnataka
3	Maharashtra Animal Science and Fishery University.	Nagpur	Maharashtra
4	Guru Angad Dev Veterinary and Animal Science University.	Ludhiana	Punjab
5	Tamil Nadu Veterinary and Animal Science University.	Chennai	Tamil Nadu
6	Tamil Nadu Fisheries University	Nagapattinam	Tamil Nadu
7	UP Pandit Deen Dayal Upadhaya Pashu Chikitsa Vigyan Vishwa Vidhyalaya evam Go Anusandhan Sansthan, Mathura	Mathura	Uttar Pradesh
8	West Bengal University of Animal and Fishery Sciences.	Kolkata	West Bengal
9	Rajasthan University of Veterinary and Animal Sciences	Bikaner	Rajasthan
10	Kerala University of Fisheries and Ocean Studies	Kochi	Kerala
11	Kerala Veterinary and Animal Sciences University	Wayanand	Kerala

III. HORTICULTURAL UNIVERSITIES

Sl.No.	Name of University	Location	State
1	Andhra Pradesh Horticultural University	Tadepalligudem	Andhra Pradesh
2	University of Horticultural Sciences	Bagalkot	Karnataka
3	Dr. Yashwant Singh Parmar Univ of Horticulture and Forestry.	Solan	Himachal Pradesh
4	Dr. YSR Horticultural University	Veketaramanna-gudem	Andhra Pradesh
5	Uttar Khand University of Horticulture and Forestry	Bharsar	Uttrakhand

Appendix 2
List of ICAR Research Institutes in India

Sl.No.	*Name of Research Institute*	*Location*	*State*
A	**CROP SCIENCES**		
1	ICAR-Central Institute for Cotton Research	Nagpur	Maharashtra
2	ICAR- Central Institute for Jute and Allied Fibres	Kolkata	West Bengal
3	ICAR-National Rice Research Institute	Cuttack	Odisha
4	Central Tobacco Research Institute	Rajamundry	Andhra Pradesh
5	ICAR- Directorate of Groundnut Research	Junagarh	Gujarat
6	ICAR-Indian Institute of Oilseeds Research	Hydereabad	Telangana
7	ICAR- Directorate of Rapeseed-Mustard Research	Bharatpur	Rajasthan
8	ICAR-Indian Institute of Rice Research	Hyderabad	Talangana
9	ICAR-Indian Institute of Seed Science	Mau	Uttar Pradesh
10	ICAR-Indian Institute of Millets Research	Hyderabad	Talangana
11	ICAR-Indian Institute of Soybean Research	Indore	Madhya Pradesh
12	ICAR-Indian Agriculture Research Institute	Pusa	New Delhi
13	ICAR-Indian Agriculture Research Institute	Hazaribagh	Jharkhand
14	ICAR-Indian Grassland and Fodder Research Institute	Jhansi	Uttar Pradesh
15	ICAR-Indian Institute of Maize Research	Ludhiana	Punjab
16	ICAR-Indian Institute of Agricultural Biotechnology	Ranchi	Bihar
17	ICAR-Indian Institute of Pulses Research	Kanpur	Uttar Pradesh
18	ICAR-Indian Institute of Sugarcane Research	Lucknow	Uttar Pradesh
19	ICAR-Indian Institute of Wheat and Barley Research	Karnal	Haryana
20	ICAR-National Bureau of Agricultural Insect Resources	Bangaluru	Karnataka
21	ICAR-National Bureau of Agriculturally Important Micro-organism	Mau	Uttar Pradesh
22	ICAR-National Bureau of Plant Genetic Resources	Pusa	New Delhi
23	ICAR-National Centre for Integrated Pest Management	Pusa	New Delhi
24	ICAR- National Institute of Biotic Stress Management	Raipur	Chhatishgarh
25	ICAR-National Centre on Plant Biotechnology	Pusa	New Delhi
26	ICAR-Sugarcane Breeding Institute	Coimbatore	Tamil Nadu
27	ICAR-Vivekanand Parvatiya Krishi Anusandhan Sansthan	Almora	Uttrakhand
B	**HORTICULTURAL SCIENCES**		
28	ICAR-Central Institute for Arid Horticulture	Bikaner	Rajasthan
29	ICAR-Central Institute for Subtropical Horticulture	Lucknow	Uttar Pradesh
30	ICAR-Central Institute for Temperate Horticulture	Srinagar	Jammu and Kashmir

Sl.No.	Name of Research Institute	Location	State
31	ICAR- Central Island Agricultural Research Institute	Portblair	Andman and Nicobar
32	ICAR-Central Plantation Crops Research Institute	Kasargod	Kerala
33	ICAR-Central Potato Research Institute	Shimla	Himachal Pradesh
34	ICAR-Central Tuber Crops Research Institute	Trivendrum	Kerala
35	ICAR-Directorate of Cashew Research	Puttur	Karnataka
36	ICAR-Directorate of Floriculture Research	Pusa	New Delhi
37	ICAR-Directorate of Medicinal And Aromatic Plants Research	Anand	Gujarat
38	ICAR-Directorate of Mushroom Research	Solan	Himachal Pradesh
39	ICAR-Directorate of Onion and Garlic Research	Pune	Maharashtra
40	ICAR-Indian Institute of Horticulture Research	Bangalore	Karnataka
41	ICAR-Indian Institute of Oil Palm Research	Pedavegi	Andhra Pradesh
42	ICAR-Indian Institute of Spices Research	Calicut	Kerala
43	ICAR-Indian Institute of Vegetable Research	Varanasi	Uttar Pradesh
44	ICAR-National Research Centre for Banana	Tiruchirapalli	Kerala
45	ICAR-Central Citrus Research Institute	Nagpur	Maharashtra
46	ICAR-National Research Centre for Grapes	Pune	Maharashtra
47	ICAR-National Research Centre for Litchi	Muzaffarpur	Bihar
48	ICAR-National Research Centre for Orchids	Gangtok	Sikkim
49	ICAR-National Research Centre on Pomegranate	Solapur	Maharashtra
C	**NATURAL RESOURCE MANAGEMENT**		
50	ICAR-National Research Centre for Seed Spices	Ajmer	Rajasthan
51	ICAR-Central Agro-forestry Research Institute	Jhansi	Uttar Pradesh
52	ICAR-Central Arid Zone Research Institute	Jodhpur	Rajasthan
53	ICAR-Central Research Institute for Dry land Agriculture	Hyderabad	Andhra Pradesh
54	ICAR-Central Soil and Water Conservation	Dehradun	Uttrakhand
55	ICAR-Central Soil Salinity Research Institute	Karnal	Haryana
56	Indian Institute of Water Management	Bhubaneswar	Odisha
57	Directorate of Weed Research	Jabalpur	Madhya Pradesh
58	ICAR Research Complex for NEH Region	Shillong	Meghalaya
59	ICAR Research Complex for Eastern Region	Patna	Bihar
60	ICAR-Indian Institute of Soil Science	Bhopal	Madhya Pradesh
61	ICAR-Indian Institute of Farming System Research	Meerut	Uttar Pradesh
62	ICAR-Indian Institute of Soil Science	Bhopal	Madhya Pradesh
63	ICAR-National Bureau of Soil Survey and Land Use Planning	Nagpur	Maharashtra
64	ICAR-National Institute of Abiotic Research management	Baramati	Maharashtra
65	ICAR- National Organic Farming Research	Gangtok	Sikkim

Sl.No.	Name of Research Institute	Location	State
66	ICAR_ National Research Centre for Integrated Farming	Morihari	Bihar
C	**AGRICULTURAL ENGINEERING**		
67	ICAR-Central Institute for Research on Cotton Technology	Mumbai	Maharashtra
68	ICAR-Central Institute of Agricultural Engineering	Bhopal	Madhya Pradesh
69	ICAR-Central Institute Post Harvest Engineering and Technology	Ludhiana	Punjab
70	ICAR-Indian Institute of Natural Resigns and Gums	Ranchi	Jharkhand
71	ICAR-National Institute of Research on Jute and Allied Fibres Technology	Kolkata	West Bengal
D	**AGRICULTURAL EDUCATION**		
72	ICAR-Central Institute for Women in Agriculture	Bhubaneswar	Odisha
73	ICAR-Indian Agricultural Statistics Research Institute	Pusa Campus	New Delhi
74	ICAR-National Academy of Agricultural Research Management	Hyderabad	Telangana
75	ICAR-National Institute of Agricultural Economics and Policy Research	Pusa Campus	New Delhi
E	**AGRICULTURAL EXTENSION**		
76	ICAR-Directorate of Knowledge Management in Agriculture	Pusa Campus	New Delhi
F	**ANIMAL RESEARCH INSTITUTES**		
77	ICAR-Central Avian Research Institute	Bareilly	Uttar Pradesh
78	ICAR-Central Institute for Research on Buffaloes	Hisar	Haryana
79	ICAR-Central Institute for Research on Cattle	Meerut	Uttar Pradesh
80	ICAR-Central Institute for Research on Goats	Mathura	Uttar Pradesh
81	ICAR-Central Sheep and Wool Research Institute	Avikanagar	Rajasthan
82	ICAR-Directorate of Poultry Research	Hyderabad	Telangana
83	ICAR-Indian Veterinary Research Institute	Bareilly	Uttar Pradesh
84	ICAR-National Bureau of Animal Genetic Resources	Karnal	Haryana
85	ICAR-National Dairy Research Institute	Karnal	Haryana
86	ICAR-National Institute of Animal Nutrition and Physiology	Karnal	Haryana
87	ICAR-National Institute of High Security Animal Diseases	Bhopal	Madhya Pradesh
88	ICAR-National Institute of Veterinary Epidemiology and Disease Informatics	Bangalore	Karnataka
89	ICAR-National Research Centre for Equines	Hisar	Haryana
90	ICAR-National Research Centre for Camel	Bikaner	Rajasthan
91	ICAR-National Research Centre on Meat	Hyderabad	Telangana
92	ICAR-National Research Centre on Mithun	Jharnapani	Nagaland
93	ICAR-National Research Centre on Pig	Guwahati	Assam

Sl.No.	Name of Research Institute	Location	State
94	ICAR-National Research Centre on yak	Dirang	Arunachal Pradesh
95	ICAR-Project Directorate on Foot and Mouth Disease	Nainital	Uttrakhand
H	**FISHERIES INSTITUTES**		
96	ICAR-Central Island Fisheries Research Institute	Barrackpur	West Bengal
97	ICAR-Central Institute of Brackish water Aquaculture	Chennai	Tamil Nadu
98	ICAR-Central Institute Fisheries Education	Mumbai	Maharashtra
99	ICAR-Central Institute Fisheries Technology	Cochin	Kerala
100	ICAR-Central Institute of Fresh Water Aquaculture	Bhubaneswar	Odisha
101	ICAR-Central Marine Fisheries Research Institute	Kochi	Kerala
102	ICAR-Directorate of Cold Water Fisheries Research	Bhimtal	Uttrakhand
103	ICAR-National Bureau of Fish Genetic Resources	Lucknow	Uttar Pradesh

Appendix 3
List of International Agricultural Research Institutes/Centres

Sl.No.	*Name of Research Institute/Centre*	*Founded in*	*Location*	*Country*
1.	International Rice Research Institute [IRRI]	1960	Manila	Philippines
2.	International Wheat and Maize Improvement Centre [CIMMYT]	1963	Mexico City	Mexico
3.	International Centre for Tropical Agriculture [CIAT]	1967	Cali	Colombia
4.	International Institute for tropical Agriculture [IITA]	1967	Ibadan	Nigeria
5.	International Potato Centre [CIP]	1971	Lima	Peru
6.	International Crop Research Institute for Semiarid Tropics [ICRISAT]	1972	Hyderabad	India
7.	Internationa Centre for Agriculture Research in Drylans Areas [ICARDA]	1977	Aleppo	Syria
8.	West Africa Rice Development Association [WARDA]	1970	Bouake	Cote divoire
9.	International Plant Genetic Resources Institute [IPGRI]	1974	Rome	Italy
10.	Centre for International Forestry Research [CIFOR]	1992	Jakarta	Indonesia
11.	International Centre for Research in Agroforestry [ICRAF]	1991	Nairobi	Kenya
12.	International Centre for Living Aquati Resources Management [ICLARM]	1977	Makati City	Philippines
13.	International Food Policy Research Institute [IFPRI]	1975	Washington DC	USA
14.	International Irrigation Management Institute [IIMI]	1984	Colombo	Sri Lanka
15.	International Livestock Research Institute [ILRI]	1995	Nairobi	Kenya
16.	International service for National Agricultural research [ISl.No.AR]	1979	Hague	Netherlands
17.	Asian Vegetable Research and Development Centre [AVRDC]	1971	Taiwan City	Taiwan [China]
18.	International Centre for Genetic Engineering and Biotechnology [ICGEB]	1987	Trieste	Italy

Appendix 4
List of Indian Seed Testing Laboratories

Sl.No.	*Name of Laboratory*	*Location/Address*	*State*
1	State Seed Testing Laboratory	Ranvir Singh Pura, Jammu	Jammu and Kashmir
2	Seed Testing Laboratory	Lalmandi, Srinagar	Jammu and Kashmir
3	State Seed Testing Laboratory	Palampur	Himachal Pradesh
4	Seed Testing Laboratory	Solan	Himachal Prades
5	Seed Testing Laboratory	PAU, Ludhiana	Punjab
6	State Seed Testing Laboratory	HAU, Hisar	Haryana
7	Seed Testing Laboratory	Karnal	Haryana
8	Quality Control Laboratory	Umri, Kurukshetra	Haryana
9	Seed Testing Laboratory	Durgapura, Jaipur	Rajasthan
10	State Seed Testing Laboratory	Agri Uni., Kanpur	Uttar Pradesh
11	Seed Testing Laboratory	Agri. Uni., Pantnagar	Uttar Khand
12	Seed Testing Laboratory	FRI, Dehradun	Uttar Khand
13	Seed Testing Laboratory	Delhi Road Meerut	Uttar Pradesh
14	Seed Testing Laboratory	Rampur Garden, Bareilly	Uttar Pradesh
15	Seed Testing Laboratory	Civil Lines, Mathura	Uttar Pradesh
16	Seed Testing Laboratory	GATDC, Jhansi	Uttar Pradesh
17	Seed Testing Laboratory	GATDC, Hardoi	Uttar Pradesh
18	Seed Testing Laboratory	GATDC, Varanasi	Uttar Pradesh
19	Seed Testing Laboratory	GATDC, Barabanki	Uttar Pradesh
20	Seed Testing Laboratory	GATDC, Azamgarh	Uttar Pradesh
21	Seed Testing Laboratory	A-264 Indira Nagar, Lucknow	Uttar Pradesh
22	Seed Testing Laboratory	Agri. Uni., Faizabad	Uttar Pradesh
23	State Seed Testing Laboratory	Gwalior	Madhya Pradesh
24	Seed Testing Laboratory	Krishi Nagar, Jabalpur	Madhya Pradesh
25	Seed Testing Laboratory	Agri. Uni., Nagpur	Maharashtra
26	Seed Testing Laboratory	Agri. College, Pune	Maharashtra
27	Seed Testing Laboratory	MAU, Parbhani	Maharashtra
28	Seed Testing Laboratory	Agri. Dept. Aurangabad	Maharashtra
29	Seed Testing Laboratory	MPKV, Rahuri	Maharashtra
30	Seed Testing Laboratory	PDKV, Akola	Maharashtra
31	State Seed Testing Laboratory	Agri. Dept. Dholi	Bihar
32	Seed Testing Laboratory	Sheikhpura, Patna	Bihar
33	Seed Testing Laboratory	RAU, Sabour	Bihar
34	Seed Testing Laboratory	BAU, Ranchi	Bihar
35	Seed Testing Laboratory	Ulubari, Gauhati	Assam

Sl.No.	Name of Laboratory	Location/Address	State
36	Seed Testing Laboratory	Shillong	Meghalaya
37	Seed Testing Laboratory	Malda	West Bengal
38	Seed Testing Laboratory	Kolkata	West Bengal
39	Seed Testing Laboratory	Agri Farm, Burdwan	West Bengal
40	Seed Testing Laboratory	Agri Dept., Ranipool	Sikkim
41	Central Seed Testing Laboratory	IARI, New Delhi	Delhi
42	Quality Control Laboratory	Pusa Campus, New Delhi	Delhi
43	Seed Testing Laboratory	Khyber Pass, New Delhi	Delhi
44	Seed Testing Laboratory	Agri. Uni., Junagarh	Gujarat
45	Seed Testing Laboratory	Agri. Dept., Navsari	Gujarat
46	Seed Testing Laboratory	Gandhi Nagar	Gujarat
47	Seed Testing Laboratory	Rajendranagar, Hyderabad	Telangana
48	Seed Testing Laboratory	Tandepalligudam	Andhra Pradesh
49	Seed Testing Laboratory	Lalaguda, Secunderabad	Telangana
50	Seed Testing Laboratory	Agri. Dept. Cuddapah	Andhra Pradesh
51	Seed Testing Laboratory	Hebbal, Bangalore	Karnataka
52	Seed Testing Laboratory	Lalbagh, Bangalore	Karnataka
53	Seed Testing Laboratory	Dharwad	Karnataka
54	Seed Testing Laboratory	Agri Dept. Bhubneshwar	Odisha
55	Seed Testing Laboratory	Agri. Dept., Panji	Goa
56	Seed Testing Laboratory	Indramanipuram, Coimbatore	Tamil Nadu
57	Seed Testing Laboratory	Tirunagar, Madurai	Tamil nadu
58	Seed Testing Laboratory	Vyalogam, Kundumian Malai	Tamil Nadu
59	Seed Testing Laboratory	ARS Pattambi, Palghat	Kerala
60	Seed Testing Laboratory	Kalar Kode, Alleppy	Kerala
61	Seed Testing Laboratory	Agri Dept. Pndicherry	Pondicherry

* * * *

Glossary

Abnormal Seedlings: Seedling that do not have capacity to develop in to normal plants. Such seedlings lack either cotyledons or primary root or have stunted root.

Absorption: The uptake of moisture into the tissues of an organism (*e.g.* seed).

Accelerated ageing: A vigor test in which the ability of seeds to germinate is assessed after their exposure to a short period of environmental stress (high temperature and humidity).

Accession: Plant material collected overseas and brought into a country for evaluation.

Achene: A small, dry, one-seeded fruit with a thin, dry wall that does not split open at maturity (*e.g.* sunflower seed).

Admixture: Something added to seed other than the species and/or cultivar specified.

Adsorption: The accumulation and adhesion of a thin layer of water (or gases) on the surface of another substance.

After-ripening: A term for collective changes that occur in a dormant seed that make it capable of germination. It is usually considered to denote physiological changes.

Agronomic Characters: Plant characteristics important in the adaptation of a plant to its commercial uses.

Air Screen Cleaner : The basic piece of equipment for cleaning seed, utilizing airflow, and perforated screens.

Aleurone Layer: The layer of high protein cells surrounding the storage cells of the endosperm. Its function is to secrete hydrolytic enzymes for digesting food reserves in the endosperm.

Angiosperm: A kind of plant that has seeds formed within an ovary.

Annual: A plant that completes it life cycle in one year or season.

Anther: The sac-like structure of the male part (stamen) of a flower in which the pollen is formed. Anthers normally have two lobes or cavities that dehisce at anthesis and allow the pollen to disperse.

Anthesis: The period of flowering, specifically the time when the stigma is ready to receive the dispersed pollen.

AOSA: The initials of the Association of Official Seed Analysts, the organization of state and federal seed analysts of the United States and Canada.

AOSCA: The initials of the Association of Official Seed Certifying Agencies, the organization of certification agencies of the United States and Canada.

Apomixis: Development of seed without sexual fusion of the egg and the sperm cells.

Aspirator: An air-blast separator. A seed conditioning (cleaning) machine that uses air to separate seeds according to specific gravity (weight) and resistance to airflow.

Asexual Reproduction: Reproduction by vegetative means without the fusion of two sexual cells.

ASTA: The initials of the American Seed Trade Association.

Awn: A slender appendage often associated with seeds, such as the "beards" of wheat or barley.

Basic Seed: A class of seed produced under the responsibility of the breeder according to the generally accepted practices for the maintenance of the cultivar which is intended for the production of Certified Seed.

Biennial: A plant that normally requires two years to complete its life cycle.

Blend: A mixture of different seed lots of the same cultivar (or species) to fulfil a specific agronomic purpose.

Breeder's Rights: Cultivar (varietal) protection; the legal rights of a breeder or owner in controlling seed production and marketing of cultivars.

Breeder's Seed: A class of seed in a seed certification program (OECD); seed produced under the supervision of the plant breeder/owner of the cultivar, which is controlled by that person or institution, and is the source of the initial and recurring increase of Basic Seed.

Broadcasting: The scattering of seeds randomly over a wide area.

Bulk: Seed in non-packaged form.

Candidate Variety: A variety that has to be protected.

Caryopsis: A dry, indehiscent one-seeded fruit (as in grasses).

Catch-cropping: Harvesting a seed crop if the season and market look favorable; an opportunistic, rather than specialist, approach to seed production.

Certified Seed: Progeny of foundation seed produced under the supervision of a seed certification agency.

Clone: A group of plants originated by vegetative propagation from a single plant.

Coleoptile: A transitory membrane covering the shoot apex of certain species that protects the plumule as it emerges through the soil.

Color Separator: A machine that separates seeds on the basis of their color differences.

Composite: In cross-pollinated species, a variety developed by mixing the seed of various genotypes which are similar in maturity, height, seed size, seed color, *etc.*

Complete Hybrid: A legal designation for a seed lot indicating that at least 95 per cent of the seed represents hybrid seed.

Conditioning: The term used to describe the process of cleaning seed and preparing it for market; also called Processing.

Conductivity Test (of seed leachates): An electrical conductivity vigor test that associates the concentration of leachates from seeds, soaking in water, to their vigor.

Cotyledon: Seed leaves of the embryo. In most dicotyledon seeds they are thickened and are storage sites of reserve food for use by the germinating seedling.

Cultivar: A cultivated variety which is clearly distinguishable by its various characters (morphological, cytological, chemical *etc.*) and which when reproduced (sexually or asexually) retains its distinguishing characters; same as variety.

Cultural Practices: All operations involved in the growing of crops, such as cultivation, sowing, spraying, irrigating, *etc.*

Day-neutral Plants: Plants that have no day-length requirement for floral initiation.

Dehumidifier: Equipment for removing moisture from the air (and therefore lowering humidity).

Dead seeds: The in-viable seeds.

Desiccant: A chemical applied to crops that prematurely kills their vegetative growth; often used for some seed crops so the seed can be harvested prior to normal plant senescence.

Dehiscence: The splitting open at maturity by pods or capsules along definite lines or sutures.

Detasselling: Artificially removing (cutting or pulling) the tassels of the female parent to prevent selfing during hybrid maize seed production.

Determinant flower: A floral axis which terminates as a flower rather than a bud.

Dicot: An abbreviated name for dicotyledonous seeds which refers to plants having two seed leaves.

Direct Heading or Harvesting: Harvesting a standing seed crop with a combine harvester without having cut the crop first.

Direct Drilling: The sowing of seeds into a soil that has not been previously cultivated to form a seedbed.

Distinctiveness: The clearly distinguishable feature of a variety from previously available varieties at least for one character.

Dormancy: A physical or physiological condition of a viable seed that prevents germination even in the presence of otherwise favorable germination conditions.

Double Cross Hybrid: The first generation progeny of a cross between two single-cross hybrids.

DUS: It refers to distinctiveness, uniformity and stability of a variety.

Ecotype: A strain within a given species adapted to a particular environment.

Electrophoresis Test: A method for separating and mapping protein bands from homogenized plant (or seed) preparations. The separations are made with a gel preparation across an electrical field. The test has potential use in cultivar identification and tests for cultivar purity.

Electrostatic Separator: A machine that separates seeds on the basis of their ability to accept and retain an electrical charge.

Embryo: The generative part of a seed that develops from the union of the egg cell and sperm cell and during germination becomes the young plant.

Emergence: The appearance of a seedling through the soil surface.

Endosperm: The tissue of seeds that develops from sexual fusion of the polar nuclei of the ovule and the second male sperm cell. It provides nutrition for the developing, growing embryo.

Epicotyl: The portion of the embryo or seedling above the cotyledons.

Epidermis: The outer layer of cells in plants that protects them against drying and mechanical injury.

Epigeal Germination: A type of germination in which the cotyledons are raised above the ground by elongation of the hypocotyl.

Ergot: Dark spur-shaped sclerotium that develops in place of a healthy seed in a diseased (fungus-infected) inflorescence. Ergot sclerotia are toxic to both man and livestock.

Example variety: A variety that is used for comparison of a particular character.

Extant variety: All released and notified varieties which have not been protected.

Excised Embryo Test: A rapid test used for evaluating the growth potential of a root-shoot axis that has been detached from the remainder of the seed.

F_1 Hybrid: A plant developed by a cross between two genetically unlike parents.

Fallowing: The practice of resting a field for several months in a year by not sowing a crop or pasture.

Farmers' Variety: A variety that has been developed by a farmer and used for commercial cultivation for several years.

Field Burning: The practice of burning the stubble and plant residues remaining after a seed crop has been harvested.

Field inspection: An inspection of a seed field, usually as a part of a seed certification program. Checks are carried out to ascertain correct crop rotation, isolation, roguing, hectarage, detasselling, male elimination, *etc.*

Field Standards: Standards of quality for the certified seed crop in the field, such as correct crop rotation, isolation distance, noxious weeds, disease level, roguing, detasselling, male elimination, *etc.*

Floral Induction: The physiological changes in response to external stimuli (light quality, day length, *etc.*) that occur in vegetative meristems and subsequently allow them to become reproductive meristems and undergo floral initiation.

Floral Initiation: The morphological changes in the development of a reproductive meristem from a vegetative meristem.

Floret: The smallest unit of a flower. In grasses it consists of the lemma, palea, stamens, and pistil.

Foundation Seed: The progeny of breeder seed; used as planting stock for registered and certified seed (USA).

Gamet Divider: A type of mechanical halving device for subdividing a large seed sample to obtain a smaller working sample for germination or purity analysis.

Genetic Drift: A change in gene and genotype frequency due to sampling error [small sample].

Genetic Purity: Trueness to type or cultivar, usually referring to seed.

Genotype: The hereditary makeup of a plant (or cultivar) which determines its inheritance).

Germination: The resumption of active growth by the embryo culminating in the development of a young plant from the seed.

Germination Test: Determination of the percentage of normal and abnormal seedlings and dead or dormant seeds in a representative sample drawn from a seed lot.

Germplasm: Hereditary properties of an individual plant or plant population that are transmitted from one generation to another.

Gravity Separator: A machine utilizing a vibrating porous deck and air flow to separate seeds on the basis of their different specific gravity.

Growth Regulator: A synthetic compound produced in the laboratory, which controls growth responses in plants and seeds.

Hard Seed: A seed which is impermeable to either water or oxygen.

Heterogeneous: Seed lots or plant populations that are not uniform.

Homogeneous: Seed lots or plant populations that are uniform.

Hybrid: The first generation seed produced by a cross between two or more genotypes.

Hybrid Vigor: The increase in performance of hybrids over that of their parental inbred types; also known as heterosis.

Hypocotyl: The part of the embryo axis between the cotyledons and the primary root which gives rise to the stalk of the young plant.

Hypogeal Germination: A type of germination in which the cotyledons remain below the ground while the epicotyl grows and emerges above the ground.

Imbibition: The initial step in seed germination involving the uptake of moisture by absorption from the germination media and hydration of the seed tissue.

Inbred Line: A plant produced by continuous selfing for several generations.

Increase (seed): To multiply a quantity of seed by planting it, thereby producing a larger quantity of seeds; also called seed increase.

Indehiscent: Not splitting open at maturity.

Indent Cylinder: A seed separator utilizing a rotating indented cylinder through which seeds are passed for cleaning. It lifts shorter seeds from longer seeds, thus separating them.

Indent Disk Separator: A seed separator utilizing multiple rotating disks inside a cylinder through which seeds are moved. It lifts shorter seeds from longer seeded types, thus separating them.

Indeterminate Flower: A plant which continues to grow through the growing season. **Inert Matter:** Nonliving materials such as sand, pebbles, soil particles, straw, *etc.* **Inflorescence:** The flowering structure of a plant (*e.g.*, umbel, spike or panicle).

Inoculum: Any material such as spores, bacteria or fungus bodies that serve as a means of propagating or spreading a pathogenic disease.

ISTA: The initials of the International Seed Testing Association.

Isolation: The separation of the field of a variety from that of another variety of the same crop to a prescribed distance to avoid contamination.

Label: An approved tag issued by a seed certifying authority, with prescribed information about the seed quality.

Laboratory Test: Analysis of seed quality in a seed testing laboratory.

Long Day Plants: Plants that initiate flowers under long day (short night) regimes.

Maintenance breeding: The process of maintaining genetic and physical purity of released and notified varieties and parents of hybrids.

Maintenance procedures: Various breeding methods used to maintain the genetic and physical purity of released and notified varieties and parents of hybrids.

Magnetic Separator: A machine that separates seeds on the basis of their magnetic characteristics (whether they are attracted or repelled by a magnet).

Male Sterile - Producing no functional pollen.

Matri-conditioning: Hydrating seeds in low water potential solid carriers such as clays or vermiculite followed by subsequent drying to enhance germination, stand establishment, and seedling growth (also known as solid matrix priming).

Model Bulk Selection: Selecting large number of single plants [300-500] from the breeder seed plot of a variety, examining these plants under laboratory conditions for the characteristics of the variety recorded at the time of its release and selecting those plants which conform to the characteristics of a variety and their seed is bulked to grow nest generation.

Monocotyledon: Group of flowering plants having one cotyledon.

Monoecious: Plants that have separate male and female flowers on the same plant.

Multiline: In self-pollinated species, a mixture of several genetically related lines.

Negative Mass Selection: Removal of undesirable plants [off types] from the field of a variety and mixing the seed of remaining plants to grow the next generation.

Nick: Synchronization of flowering between male and female inbred lines during hybrid seed production, by adjusting planting times.

Nitrogen Fixation: The ability of legumes to utilize N2 in the atmosphere through an association with soil-borne Rhizobium bacteria.

Normal seedlings: Seedlings with normal growth of root and shoot.

Novelty: The newness of a variety. A variety which has not been grown for more than one year prior to application of its registration.

Noxious Weed: A highly objectionable weed found in seed lots.

Nucleus seed: The initial seed of an improved variety, limited in quantity and produced by originating plant breeder.

OECD: Organization for Economic Co-operation and Development; international agency which, amongst other things, runs an international seed certification scheme.

Off-type: A plant that deviates in one or more characteristics from the improved cultivar under multiplication.

Open Pollinated Cultivar: A heterogeneous cultivar of a cross-pollinated crop that is allowed to inter-pollinate freely during seed production.

Osmo-conditioning: Soaking seeds in water or salts followed by subsequent drying to enhance germination, stand establishment, and seedling growth (also known as priming).

Other Crop Seed: One of the components of a purity test; the total percentage (by weight) of seed of all crop species.

Ottawa Seed Blower: A type of seed blower developed by C.W. Leggitt of the Canadian Department of Agriculture that has a slender metal blowing tube used for small seeded crops.

Ovary: The part of the pistil containing the ovule.

Ovule: The structure within the ovary of the flower that becomes the seed following fertilization and development.

Panicle: An inflorescence in which the lateral branches arising from the peduncle produce flower bearing branches instead of single flowers.

Parental Material: A limited number of plants selected and used for the maintenance of a cultivar.

Pathogen: A disease causing organism.

Pelleted Seeds: The process of coating seeds with special nutrients or growth-promoting substances to aid in seed germination and growth.

Perennial: A plant which survives and produces vegetative growth and flowers year after year without being replanted.

Physical purity: Freedom of seed from inert matter and defective seeds.

Physiological Dormancy: Seed dormancy caused by internal physiological conditions that prevent germination; often referred to as epicotyl or embryo dormancy.

Physiological Maturity: The maturity of a seed when it reaches its maximum dry weight. This usually occurs prior to the normal harvest date.

Phytosanitary Certificate: A certificate issued by a legally constituted authority of government stating that a seed lot has been inspected and found to be free of disease infestation. These certificates are frequently used in international seed trade agreements to prevent the spread of seed-borne diseases among countries.

Pistil: The female, or seed-bearing organ of the flower, composed of the ovary, style and stigma.

Plant to Row Method: This method of maintenance breeding is based on progeny test. It is also known as progeny row selection. Progeny of single plant selections are planted, evaluated and progeny with true to the type are selected and their seed is mixed to grow the nest generation.

Pollen: The small, almost microscopic, yellow bodies that are borne within the anthers of flowers and contain the male generative (sex) cells.

Pollination: The process by which pollen is transferred from the anther where it is produced to the stigma of a flower.

Population: A group of plants of the same species kept in a group to achieve certain plant breeding objectives.

Positive Mass Selection: Selection of desirable plants [true to type] from the field of a variety and mixing the seed of individual selected plants to grow the next generation.

Pre-chill: The practice of exposing imbibed seeds to cool (5-10°C) temperature conditions for a few days prior to germination at warmer conditions. See definition of stratification.

Pre-hydration: Soaking seeds in water or gels prior to planting to enhance germination, stand establishment, and seedling growth.

Priming: Soaking seeds in water or salts followed by subsequent drying to enhance germination, stand establishment, and seedling growth (also known as osmo onditioning).

Processing: The term used to describe the technology of cleaning seed and preparing it for the market. See also conditioning.

Progeny: Offspring.

Propagule: Any type of plant part used to reproduce another individual plant asexually.

Pure Line: Progeny of a self-fertilized plant obtained by selfing.

Pure Live Seed (PLS): The percentage of pure seeds in a seed lot that have the ability to germinate. The percentage of PLS is determined by multiplying percent germination by the percent pure seed and dividing by 100.

Pure Seed Content: The percentage of the species stated by the sender or found to be predominant in the testing of a seed lot.

Purity Physical: The percentage by weight of pure seed of a species as determined by a seed-testing laboratory.

Purity Genetic: The purity with respect to cultivar as determined by a field inspection, by special laboratory tests where possible, or by inspection of plots for this purpose.

Radicle: The rudimentary root of the seed or seedling that forms the primary root of the young plant.

Registered seed: Progeny of either foundation or registered seed. In India, this category is generally omitted and certified seed is produced directly from foundation seed.

Relative Humidity: Ratio of the quantity of water vapor actually in the air to the greatest amount possible at a given temperature.

Reference variety: All released and notified extant varieties that are under seed production chain.

Rogue: An off-type plant in the field of an improved variety.

Roguing: The process of removal of the off types plants from the field of an improved variety to avoid contamination.

Sampling: The method by which a representative sample is taken from a seed lot to be sent to a laboratory for analysis.

Scalper: A screening or air blast machine used to separate the very large, very small, or very light contamination from a seed lot - a rough cleaner.

Scarification: The process of mechanically abrading a seed coat to make it more permeable to water. This process may also be accomplished by brief exposure to strong acids (*e.g.*, sulphuric acid).

SCST: The initials of the Society of Commercial Seed Technologists, an organization of commercial and private registered seed technologists and seed analysts of the United States and Canada.

Secondary Dormancy: A type of dormancy imposed by certain adverse environmental conditions in previously non-dormant seeds, or seeds in which primary dormancy have been broken.

Seed: A mature ovule consisting of an embryonic plant together with a store of food, all surrounded by a protective coat.

Seedbed: Soil that has been cultivated to provide conditions suitable for seed germination, emergence and growth.

Seed-borne: Carried on or in seeds.

Seed certification: A legal seed system that ensures availability of high quality seed of improved variety to the farmers.

Seed Coat: The protective covering of a seed usually composed of the inner and outer integuments; also called the testa.

Seed Coating: Application of substances such as fungicides, insecticides, micronutrients, *etc.* directly to the seed that do not obscure its shape.

Seed Enterprise: Any organization involved in seed growing, either directly or through contracts with others, drying, processing, storage, and marketing.

Seed grower: An individual or institution that grows seed but does not process or market seed.

Seed Increase: The multiplication of a quantity of seed by planting it to produce a larger quantity of seed.

Seed Industry: The entire complex of organizations, institutions, and individuals associated with the seed program of a country. Seed industry is involved in producing and marketing seeds for sale to consumers.

Seed lot: A quantity of seed identified by a number or other mark, every portion or bag of which is uniform within recognized tolerances and the factors that are specified or appear on a label and/or seed analysis certificate.

Seedling: A young plant grown from seed.

Seed Marketing: The systematic determination of consumer needs, accumulation of seed and services to satisfy those needs, communication of information about the availability of seed and services, and distribution of seed to consumers.

Seed Multiplier or Producer: An individual who not only grows seed but also may process and sell seed, usually on a limited scale.

Seed Multiplication Factor: The number of kg seed harvested for each kg sown.

Seed Program: The measures to be implemented and activities being carried out in a country to achieve the timely production and supply of seed of a prescribed quality in the quantities needed.

Seed Set: Commencement of early growth of the embryo and endosperm.

Seed stock: Seed that is used for multiplying additional cycles or generations of a cultivar.

Seed testing: The process of seed analysis in terms of its physical purity and germination percentage.

Seed vigor: The ability of seed to cope up with environmental stresses after sowing and/or storage.

Short Day Plants: Plants that initiate flowers best under short day (long night) regimes.

Single Cross Hybrid: A hybrid between two inbred lines.

Solid Matrix Priming: Hydrating seeds in low water potential solid carriers such as clays or vermiculite followed by subsequent drying to enhance germination, stand establishment, and seedling growth (also known as matri-conditioning).

Spike: A basic type of inflorescence in which the flowers arising along the rachis are essentially sessile (stalk-less).

Spikelet: The unit of the grass flower that includes the two basal glumes subtending one to several florets.

Spiral separator: A type of seed separators which has no moving parts. The seeds enter at the top and slide or roll down an inclined spiral runway. The speed of seed movement and centrifugal force allows the separation of heavier, round, fast-moving seeds from those that move slower.

Stability: The same performance of a variety after repeated reproduction or propagation

Stamen: The part of the flower bearing the male reproductive cells composed of the anthers on the filament (stalk).

Stigma: The upper part of the pistil that receives the pollen.

Stratification: The practice of exposing imbibed seeds to cool (5-10°C) temperature conditions for a few days prior to germination to break dormancy. This is a standard practice in germination testing of many grasses and woody species.

Strain: A term sometimes used to designate an improved selection within a cultivar.

Sub-sampling: The procedure (usually by halving methods) by which a smaller representative working sample is obtained from the larger sample submitted for seed analysis.

Surface Sowing: Sowing seeds directly onto the soil surface without subsequent soil cover.

Swath: A row of cut or pulled crop usually waiting for drying before harvesting.

Syn-1 Syn-2: The first and second-generation progenies, respectively, of a synthetic cultivar after the individual lines are composited.

Synthetic Seeds: Seeds produced from vegetative tissues (usually by tissue culture) that are clones possessing identical genotypes.

Synthetic Cultivars: A cultivar composed of an inbreeding population of several cross-pollinated plant lines.

Testa: The outer covering of the seed; the seed coat.

Tetrazolium (TZ): Indicates a class of chemicals that have the ability to accept hydrogen atoms (and undergo reduction) from dehydrogenase enzymes during the respiration process in viable seeds. This is the basis of the tetrazolium test during which the tetrazolium chemical undergoes a color change, usually from colourless to red.

Tetrazolium Test: Quick test to determine seed viability.

Three-way Hybrid: A hybrid between an inbred line and single cross hybrid.

Tiller: A branch arising from the base of a monocot plant, especially in the grass family.

Trier: A hand manipulated probe for sampling seeds.

Uniformity: The same type of population of a variety.

UPOV: International Union for the Protection of New Varieties of Plants.

Variety: A genotype released for commercial cultivation either by State Variety Release Committee or Central Variety Release Committee.

Varietal deterioration: Permanent reduction either in the genetic or agronomic value of a released variety.

Vegetative: Refers to asexual (stem, leaf, root) development in plants in contrast to sexual (flower, seed) development.

Vernalization: The process of promoting floral induction through chilling treatment.

Viable (Viability): The capacity of seed to germinate under favorable conditions in the absence of dormancy.

Volunteer Plants: Unwanted plants growing from residual seeds of previous crops.

Weed: Any plant in a place where it is a nuisance might be considered a weed.

Weed Seed (Percent): The total percentage (by weight) of a seed lot, which is composed of seed of plants, considered to be weeds.

Windrow: A loose, continuous row of cut or uprooted plants usually allowed to dry in place until harvested; a swath.

* * * *

References

1. **ALLARD, R. W.** 1960. Principles of Plant Breeding. John Wiley and Sons Inc. New York.
2. **ALLARD, R. W.** 1999. Principles of Plant Breeding 2nd edition. John Wiley and Sons Inc. New York.
3. **BOROJEVIC, S.** 1990 Principles and Methods of Plant Breeding. Elsevier, Amsterdam.
4. **BRIGGS, F. N. AND KNOWLES, P. F.** 1967. Introduction to Plant Breeding. Reinhold, New York.
5. **CHOPRA, V. L. (ED.)** 2000. Plant Breeding: Theory and Practice 2nd edition, Oxford and IBH Publishing Company, Pvt. Ltd. New Delhi.
6. **FEHR, W. R.** 1987. Principles of Cultivar Development Volume 1. Theory and Techniques. Macmillan Publishing Company, New York.
7. **FREY, K. J.** (ed.) 1966. Plant Breeding. Iowa State University, Press, Ames, Iowa, USA.
8. **FREY, K. J.** (ed.) 1981. Plant Breeding II. Iowa State University, Press, Ames, Iowa, USA.
9. **HARTEN, A. V.** 1998. Mutation Breeding: Theory and Practical Applications. Cambridge Uni. UK.
10. **KUCKKUCK, H., KOBADE, G. AND WENGEL G.** 1993. Fundamentals of Plant Breeding. Narosa Publishing House, New Delhi.
11. **MAYO, O. 1984**. The Theory of Plant Breeding. Clarendon, Oxford.
12. **POEHLMAN, J. M.** 1987. Breeding Field Crops 3rd edn.AVI Publishing Co. Inc. West Port, Connecticut, USA.
13. **POEHLMAN, J. M. AND BORTHAKUR, D. N.** 1969. Breeding Asian Field Crops. Oxford and IBH Publishing Company, New Delhi.

14. **PHUNDAN SINGH** 2018. Essentials of Plant Breeding 7th edition, Kalyani Publishers, New Delhi.

15. **PHUNDAN SINGH** 2018. Plant Breeding: Molecular and New Approaches 4th edition, Kalyani Publishers, New Delhi.

16. **PHUNDAN SINGH** 2019. Elements of Genetics 5th edition, Kalyani Publishers, New Delhi.

17. **PHUNDAN SINGH** 2012. Cotton Breeding 3rd edition, Kalyani Publishers, New Delhi.

18. **PHUNDAN SINGH AND SANJEEV SINGH** 2010. Breeding Hybrid Cotton 3rd edition, Kalyani Publishers, New Delhi.

19. **SHARMA, J. R.** 1984. Principles and Practice of Plant Breeding. Tata McGraw-Hill Publishing Company Ltd. New Delhi.

20. **SIMMONDS, N. W.** 1979. Principles of Crop Improvement. Longman, London.

21. **SNEEP, J. AND HENDRIKSON, A. J. T.** (eds.) 1979. Plant Breeding Perspectives. Pudoc, Wageningen, Netherlands.

22. **VOSE, P. B. AND BLIXT, S. G.** (eds.) 1984. Crop Breeding: A Contemporary Basis. Pergamon Press, London.

23. **WILLIAMS, W.** 1964. Genetic principles and Plant Breeding, Oxford, Blackwell.

* * * *

Books by the Senior Author

A. Plant Breeding

1. Essentials of Plant Breeding
2. Fundamentals of Plant Breeding
3. Plant Breeding: Molecular and New Approaches
4. Molecular Plant Breeding
5. Principles of Seed Technology
6. Seed Technology: At a Glance
7. Objective Seed Technology
8. Practical in Crop Breeding
9. Practical and Numerical Problems in Plant Breeding
10. Numerical Problems in Plant Breeding and Genetics
11. Objective Science of Plant Breeding
12. Plant Breeding At a Glance
13. Plant Breeding [For Under Graduate Students]
14. Principles of Plant Breeding
15. Cotton Breeding
16. Heterosis Breeding in Cotton
17. Breeding Hybrid Cotton
18. Breeding Transgenic Bt. Cotton
19. Cotton Improvement in India
20. Sankar Kapas [In Hindi]
21. Glimpses of Cotton Breeding
22. Glossary of Plant Breeding and Genetics

23. Elements of Baby Corn
24. Breeding Crop Plants for Stress Resistance
25. Plant Breeding: Related Legislations
26. Concepts in Plant Breeding

B. Genetics

27. Elements of Genetics
28. Genetics
29. Principles of Genetics [For UG Students]
30. Plant Genetics
31. Fundamentals of Genetics
32. Cotton Genetics
33. Objective Genetics
34. Genetics At a Glance
35. Objective Genetics and Plant Breeding
36. Objective Genetics and Plant Breeding [Hindi Edition]
37. Glossary cum Dictionary of Genetics and Plant Breeding
38. Molecular Genetics [Objective]
39. Molecular Genetics [At a Glance]
40. Molecular Genetics [Subjective]
41. Genetics and Man
42. Genetics Today

C. Quantitative Genetics

43. Biometrical Techniques in Plant Breeding
44. Application of Biometrical Techniques in Plant Breeding [Hindi Edition]
45. Objective Quantitative Genetics
46. Quantitative Genetics At a Glance
47. Quantitative Genetics

D. Plant Biotechnology

48. Introduction to Biotechnology
49. Plant Biotechnology
50. Principles of Plant Biotechnology
51. Plant Biotechnology: At a Glance
52. Objective Plant Biotechnology

E. Intellectual Property Rights

53. IPR and Plant Breeders' Rights [Subjective]
54. IPR and Plant Breeders' Rights [Objective]
55. IPR and Plant Breeders' Rights [At a Glance]
56. Introduction to Intellectual Property Rights
57. Intellectual Property Rights At A Glance
58. Intellectual Property Rights: Objective

F. Competitive Examination

59. A Guide to Competitive Examinations of Agriculture
60. Objective General Agriculture for Competitive Examinations

* * * *

Index

F

G

H

I

J

L

M

N